T0215778

Dynamics and Control of DC-DC Converters

Synthesis Lectures on Power Electronics Series

Editor

Jerry Hudgins, *University of Nebraska, Lincoln*

Synthesis Lectures on Power Electronics will publish 50- to 100-page publications on topics related to power electronics, ancillary components, packaging and integration, electric machines and their drive systems, as well as related subjects such as EMI and power quality. Each lecture develops a particular topic with the requisite introductory material and progresses to more advanced subject matter such that a comprehensive body of knowledge is encompassed. Simulation and modeling techniques and examples are included where applicable. The authors selected to write the lectures are leading experts on each subject who have extensive backgrounds in the theory, design, and implementation of power electronics, and electric machines and drives.

The series is designed to meet the demands of modern engineers, technologists, and engineering managers who face the increased electrification and proliferation of power processing systems into all aspects of electrical engineering applications and must learn to design, incorporate, or maintain these systems.

Modeling Bipolar Power Semiconductor Devices
Tanya K. Gachovska, Jerry L. Hudgins, Enrico Santi, Angus Bryant, and Patrick R. Palmer
2013

Signal Processing for Solar Array Monitoring, Fault Detection, and Optimization
Mahesh Banavar, Henry Braun, Santoshi Tejasri Buddha, Venkatachalam Krishnan, Andreas
Spanias, Shinichi Takada, Toru Takehara, Cihan Tepedelenlioglu, and Ted Yeider
2012

The Smart Grid: Adapting the Power System to New Challenges
Math H.J. Bollen
2011

Digital Control in Power Electronics
Simone Buso and Paolo Mattavelli
2006

Power Electronics for Modern Wind Turbines
Frede Blaabjerg and Zhe Chen
2006

© Springer Nature Switzerland AG 2022

Reprint of original edition © Morgan & Claypool 2018

Dynamics and Control of DC-DC Converters

Farzin Asadi and Kei Eguchi

ISBN: 978-3-031-01374-4   paperback
ISBN: 978-3-031-02502-0   ebook
ISBN: 978-3-031-00323-3   hardcover

DOI 10.1007/978-3-031-02502-0

A Publication in the Springer series
SYNTHESIS LECTURES ON POWER ELECTRONICS SERIES

Lecture #10
Series ISSN
Print 1931-9525   Electronic 1931-9533

Dynamics and Control of DC-DC Converters

Farzin Asadi
Kocaeli University, Turkey

Kei Eguchi
Fukuoka Institute of Technology, Japan

SYNTHESIS LECTURES ON POWER ELECTRONICS SERIES #10

ABSTRACT

DC-DC converters have many applications in the modern world. They provide the required power to the communication backbones, they are used in digital devices like laptops and cell phones, and they have widespread applications in electric cars, to just name a few.

DC-DC converters require negative feedback to provide a suitable output voltage or current for the load. Obtaining a stable output voltage or current in presence of disturbances such as: input voltage changes and/or output load changes seems impossible without some form of control.

This book tries to train the art of controller design for DC-DC converters. Chapter 1 introduces the DC-DC converters briefly. It is assumed that the reader has the basic knowledge of DC-DC converter (i.e., a basic course in power electronics).

The reader learns the disadvantages of open loop control in Chapter 2. Simulation of DC-DC converters with the aid of Simulink®is discussed in this chapter as well. Extracting the dynamic models of DC-DC converters is studied in Chapter 3. We show how MATLAB®and a software named KUCA can be used to do the cumbersome and error-prone process of modeling automatically. Obtaining the transfer functions using PSIM®is studied as well.

These days, softwares are an integral part of engineering sciences. Control engineering is not an exception by any means. Keeping this in mind, we design the controllers using MAT-LAB®in Chapter 4.

Finally, references are provided at the end of each chapter to suggest more information for an interested reader. The intended audiencies for this book are practice engineers and academi-ans.

KEYWORDS

control of DC-DC converters, dynamics of DC-DC converters, loop shaping, PID control of DC-DC converters, state space averaging, system identification, modeling of power electronics converters

*We dedicate this book to our parents
and our lovely families.*

Contents

Preface

DC-DC converters are an integral part of our modern life. They convert a voltage level to another with high efficiency.

DC-DC converters are nonlinear variable structure systems. They are subject to disturbances such as input voltage changes and output load changes. Obtaining stable output voltage seems impossible without some form of feedback control. This book help you design the control loop for DC-DC converters in a practical manner.

Although control engineering has made tremendous progress in the last decade, about 90% of applications use proportional-integral-derivative (PID) controllers. DC-DC converters are not an exception by any means and PID controllers are good enough for most common DC-DC converters. PID controller design for DC-DC converters is studied in Chapter 4.

This book discusses the dynamics and control of DC-DC converters. We assume that the reader already knows the basics of DC-DC converters and linear control theory. There are plenty of textbooks available on power electronics and linear control and one can refer to the references at the end of first chapter if a review of concepts is needed.

A brief summary of the book chapters is as follows:

Chapter 1 is a brief introduction to the world of DC-DC converters. Some of the applications of DC-DC converters are introduced in this chapter.

Chapter 2 describes the importance of control in DC-DC converters. Some simulations show what happens if the system works without any controller (i.e., open loop). This chapter also introduces the simulation of power electronics circuits with the aid of the Simscape library of Simulink®.

Chapter 3 describes the modeling procedure for DC-DC converters. DC-DC converters are nonlinear variable structure systems by nature. There are some methods available in the literature to obtain a Linear Time Invariant (LTI) small signal model of the converter. We used State Space Averaging (SSA) in this chapter to model converters working in Continuous Current Mode (CCM). A software developed by the first author is introduced to do the SSA procedure automatically. Using this software you can do the modeling automatically without any hand calculation. Computer methods of obtaining the converter frequency response is introduced in this chapter as well. A method to model converters working in DCM is introduced at the end of this chapter.

Chapter 4 describes the controller design procedure for DC-DC converters. These days, software and especially MATLAB®are an important part of control engineering. We use the MATLAB's "Control System Toolbox™" extensively in this book. Using this toolbox, a controller can be designed easily even by a novice.

We hope that this book will be useful to the readers, and we welcome comments on the book.

Farzin Asadi and Kei Eguchi
farzin.asadi@kocaeli.edu.tr   eguti@fit.ac.jp
February 2018

DC-DC Converters: An Introduction

1.1 INTRODUCTION

In this chapter we review the DC-DC converters very briefly. If you need more information on the subject see the references at the end of the chapter.

1.2 SWITCHING DC-DC CONVERTERS

Power electronics is the application of solid-state electronics to the control and conversion of electric power. Power electronics converters can be divided into four groups based on their input and output:

- AC-DC (rectifier),

- DC-AC (inverter),

- DC-DC (DC-DC converter), and

- AC-AC (AC-AC converter).

This book is involved with DC-DC converters. A DC-DC converter is a circuit to convert a source of Direct Current (DC) from one voltage level to another. DC-DC converters are an integral part of modern life. Here is a quick review of DC-DC converters. Assume that we want to supply a load of 1 Ω with the aid of a 9 V battery. The load requires 3 V so input voltage must be decreased. If we place a resistor of 2 Ω in a series with the load we provide the required voltage.

We can calculate the circuit efficiency easily as follows:

$$I = \frac{9}{2+1} = 3 \text{ A} \tag{1.1}$$

$$P_{IN} = 9 \text{ V} \times 3 \text{ A} = 27 \text{ W} \tag{1.2}$$

$$P_{Load} = 1 \text{ Ω} \times (3 \text{ A})^2 = 9 \text{ W} \tag{1.3}$$

$$\eta = \frac{9}{27} = 33\%. \tag{1.4}$$

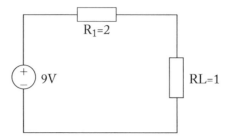

Figure 1.1: A resistive divider.

Therefore, only 9 W of input power is consumed by the load. The remaining part $(27 - 9 = 18$ W) is converted into heat.

Instead of the resistor R_1 in Fig. 1.1, one may use a transistor like that shown in Figs. 1.2 or 1.3. Zener breakdown voltage is about 3.7 V so load voltage is $V_Z - V_{BE,on} \cong 3.7 - 0.7 = 3$ V.

The efficiency of *linear* regulators shown in Figs. 1.2 and 1.3 is the same as the resistive divider shown in Fig. 1.1. The collector-emitter voltage is 6 V since the collector is connected to a 9 V supply and the emitter is connected to the load with a voltage of 3 V. The current passing from the transistor is the same as the load current which is 3 A. So, the transistor dissipates $V_{CE} \times I_C = 6$ V $\times 3$ A $= 18$ W and the efficiency is not increased in this case as well. Only 33%!

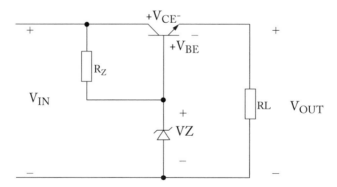

Figure 1.2: A linear voltage regulator.

As an attempt to obtain a more efficient solution, one may suggest the circuit shown in Fig. 1.4.

In this circuit, the switch S is assumed to be ideal (i.e., act like short circuit when it is closed and act like open circuit when it is opened). We place the key in position 1 or 2 with the aid of a control signal. When the key is in the position labeled "1", the output voltage is 9 V and

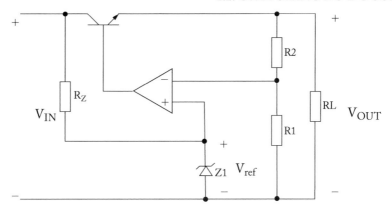

Figure 1.3: A linear voltage regulator with feedback control.

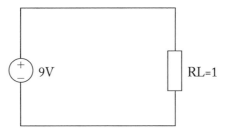

Figure 1.4: Basic idea of a switching converter.

when it is in the position "2" the output voltage is 0 V. Equivalent circuit for position 1 and 2 is shown Figs. 1.5 and 1.6, respectively.

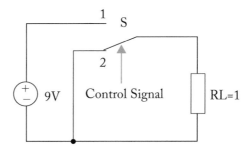

Figure 1.5: Equivalent circuit for the switch in position 1.

If we apply the control signal shown in Fig. 1.7, the voltage shown in Fig. 1.8 is obtained. T is the switching period and is in the ms or μs range.

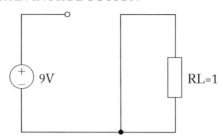

Figure 1.6: Equivalent circuit for the switch in position 2.

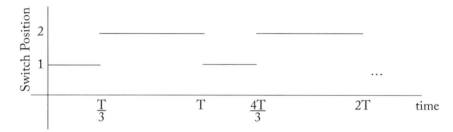

Figure 1.7: Switching pattern to obtain average output of 3 V.

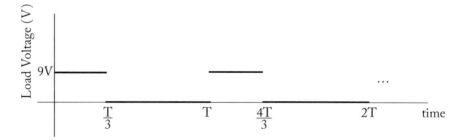

Figure 1.8: Output voltage for the switching pattern shown in Fig. 1.7.

In this case, the average value of output voltage is 3 V since

$$V_{O,ave} = \frac{1}{T}\int_0^{\frac{T}{3}} 9\,dt = \frac{1}{T}\times 9 \times \frac{T}{3} = 3 \text{ V}. \qquad (1.5)$$

So we obtain the 3 V although it is not pure DC. If we use some form of low-pass filter we can get rid of harmonics available in the output voltage.

The efficiency of this circuit is 100% since the switch is assumed to be ideal. When the switch is in the position numbered "1", the current pass from the switch is 9 A. Since the switch resistance is assumed to be 0, the dissipated power is $P_{diss,1} = RI^2 = 0 \times 9^2 = 0$ W. When the switch is in position number 2, no current passes from the switch so in this case the dissipated power ($P_{diss,2}$) is 0, too.

In a real switch, there is some form of dissipated power. Despite these dissipations, the efficiency of the conversion is high enough. So, the first benefit behind using the switching converters is an increase in the efficiency.

Figures 1.9 and 1.10 show a linear laboratory power supply and a laptop adaptor, respectively.

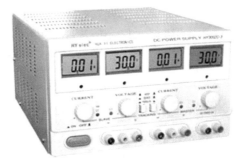

Figure 1.9: A linear laboratory power supply.

Figure 1.10: Laptop adaptor.

The linear laboratory power supply weighs about 9 kg and provides 30 V and maximum output current of 3 A. So, the maximum output power is 30 V × 3 A = 90 W. The efficiency is about 40%.

The laptop adaptor weights about 400 g and provides 19 V and maximum output current of 4.74 A. So, the given maximum output power is 19 V × 4.74 A = 90.06 W. The efficiency is about 85%.

Although the output powers are the same, the weights and volumes are not. Generally, the switching converter are smaller and more compact than their linear versions. The switching converters works in a high frequency. This makes it possible to use components with lower values (i.e., smaller capacitor and inductors). Smaller values lead to smaller components which decrease the overall volume of the converter.

A question may rise here: "Why aren't linear power supplies used in laboratories yet?" Although the switching power supplies are superior in terms of compactness and efficiency, they are not as good as linear power supplies in terms of noise. Switching power supplies contain switches which work at high-frequency. The produced high-frequency noise can reach[1] sensitive measurement devices available in the laboratory and cause error. So, it is better to keep such a noise source away from sensitive measurement devices.

1.3   SOME APPLICATIONS OF DC-DC CONVERTERS

In this section some of the applications of DC-DC converters are introduced. DC-DC converter applications are not limited to the ones written below. In fact, listing all the applications of DC-DC converters would require a separate book.

1.4   APPLICATION IN RENEWABLE ENERGY SYSTEMS

These days, renewable energies gained a lot of attention. Conversion of sun energy to electric energy can be done with the aid of a solar panel.

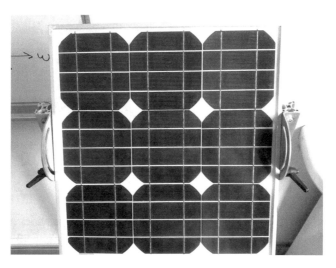

Figure 1.11: A solar panel.

Normally, the output of solar panels is a low level voltage. In order to obtain a higher voltage required by most loads, a boost converter must be used. Obtaining the maximum power from a panel is done with the aid of a DC-DC converter as well.

---

[1]The grid or electromagnetic radiations can aid the noise to reach another device.

1.5 DIGITAL DEVICES

DC-DC converters are used in digital devices such as laptops, cell phones, PDAs, etc. For example, a laptop battery may have a voltage as high as 10.8 V but the CPU requires only 3.3 V or even lower voltages to operate. So, a DC-DC converter is needed to reduce the input voltage.

1.6 CHARGING BATTERIES

Assume that you are traveling and you need to charge your cell phone using your car's battery. The car battery provides about 12 V while charging the phone battery requires only about 5 V. You cannot connect the phone's battery directly to the the car's battery, so you need to use a DC-DC converter to produce 5 V from the car battery.

Figure 1.12: A step-down converter to charge cell phones using car batteries.

You can even charge your phone using a 1.5 V battery. In this case, you use a DC-DC converter to step up the voltage to 5 V.

1.7 ANATOMY OF POWER ELECTRONICS CONVERTERS

The anatomy of power electronics converters is shown in Fig. 1.14.
As you can see, all power electronics converters are composed of four components:

- Energy Source,

- Power Circuit,

- Control Circuit, and

- Load.

Figure 1.13: A step-up converter to charge cell phones using 1.5 V batteries.

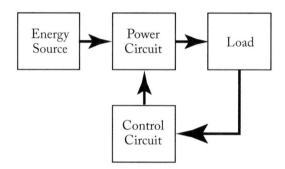

Figure 1.14: Block diagram of a power electronics converter using closed-loop control.

The "energy source" provides the required energy for the load. Sometimes it is not possible to connect the load directly to the energy source. For example the energy source may be batteries while the load requires AC voltage (i.e., AC motors). As another example, an input energy source may be a low voltage DC source while the load requires a high DC voltage.

The "power circuit" is composed of switches, inductors, and capacitors. It does the conversion of energy so the output is in the form required by the load. The power circuit is similar to a body's muscles.

The "control circuit" may be considered as the brain of system. It provides the required signals to "power circuit."

"Control circuit" can be divided into two broad categories. The first category uses some form of feedback taken from load voltage/current and produces the control signal based on the measurements. The schematic of this type of converters is shown in Fig. 1.14.

The second category uses no feedback. In other words, there is no sensor to sense the load voltage/current. A designer sets the control circuit based on nominal values and *hopes* that output is what required by the load. If the operating condition changes, output may not be what is required by the load. The schematic of this type of converter is shown in Fig. 1.15.

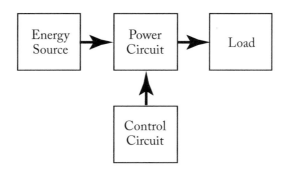

Figure 1.15: Block diagram of a power electronics converter using open-loop control.

For example, assume that we want to design a portable cell phone charger. A designer may assume that the car battery has an average voltage of 12 V and set the controller based on 12 V input. The user may use the charger in tracks or buses which usually have 24 V electric system. In this case, the charger provides an output voltage that is about two times greater than what is required. There is no mechanism to inform the control circuit about the increase in output voltage.

Needles to say, converters which use a closed-loop control system are more accurate and more expensive than converters that use open-loop control.

1.8   ABOUT THIS BOOK

The follow is a brief summary of upcoming chapters.

Chapter 2 shows you why control is important in DC-DC converters. This chapter also introduces you to the required steps to simulate a DC-DC converter in Simulink®environment.

Chapter 3 describes the modeling procedure for DC-DC converters. In this chapter we introduce a method named State Space Averaging (SSA) to extract an Linear Time Invariant (LTI) model for the converter. We introduce a software tool to extract the converter model automatically. Computer methods of extracting the converter dynamics are discussed as well.

Chapter 4 designs a controller based on models obtained from Chapter 3. Special emphasis is on the proportional-integral-derivative (PID) control since it's simple, cheap, and works well enough for most applications. More advanced methods like loop shaping, Linear Quadratic Gaussian (LQG), and Internal Model Control (IMC) can be used if you prefer to use them. We design the controllers with the aid of powerful MATLAB®tools. If you prefer hand calculations, you can refer to some of the traditional books introduced in the references.

1.9 CONCLUSION

In this chapter we introduced DC-DC converters, some of their applications, and the general block diagram of a power electronics converter. We introduced the benefits behind closed-loop control of DC-DC converters. The next chapter demonstrates, in more detail, what happens when a DC-DC converter works without any closed-loop controller in.

REFERENCES

[1] Daniel Hart, *Power Electronics*, McGraw Hill, 2011.

[2] Ned Mohan and Tore Undeland, *Power Electronics Converters, Applications and Design*, Wiley, 2002.

[3] Muhammad Rashid, *Power Electronics Devices, Circuits and Applications*, Pearson, 2013.

[4] Robert Erikson and Dragan Maksimovic, *Fundamentals of Power Electronics*, Springer, 2001. DOI: 10.1007/b100747.

[5] Simon Ang and Alejandro Oliva, *Power Switching Converters*, Taylor & Francis, 2005.

[6] Marian K. Kazimierczuck, *Pulse Width Modulated DC-DC Power Converters*, John Wiley, 2012. DOI: 10.1002/9780470694640.

[7] Katsuhiko Ogata, *Matlab for Control Engineers*, Pearson, 2007.

[8] Brian Hahn and Daniel T. Valentine, *Essential Matlab for Engineers and Scientists*, Academic Press, 2016.

[9] Holly More and Somitra Kumar Sanadhya, *Matlab for Engineers*, Pearson, 2014.

CHAPTER 2

Importance of Control in DC-DC Converters

2.1 INTRODUCTION

DC-DC converters are dynamical systems which are subject to some disturbances. The most common disturbances are input voltage changes and output load changes. We want to keep the output voltage constant in spite of disturbances.

An example is quite helpful. Assume a digital device such as a laptop or cell phone. When you use the device the battery energy is discharged and its voltage decreases with time. The device's CPU needs a constant ripple-free voltage to operate. If we use an open-loop system (i.e., no feedback controller), the changes in input voltage will affect the output voltage so the CPU voltage changes as the battery discharges.

When you do some processing with your laptop, the CPU consumes more power than when in idle mode. So, the output load (in this example the CPU) changes based on what you are doing. In a system without any controller, the output voltage in idle mode is higher (since it uses less current) than full process mode. The CPU will not work well with such an input voltage.

In this chapter, we will show you the importance of control in DC-DC converters with the aid of simulations. We do the simulations in Simulink®environment and show the required steps to set up a simulation in Simulink®environment. We encourage you to do the simulations and see the results. Simulation, especially in a user-friendly environment such as Simulink®, can help you learn the concepts more easily.

2.2 SIMULATION OF DC-DC CONVERTERS IN Simulink®

Simulation of power electronics converters can be done with the aid of Simulink's Simscape library. In order to enter Simulink®, "simulink" must be written in the MATLAB®command line:

After pressing the Enter key, a Window is opened (Fig. 2.2). We click the "Blank Model" icon.

Simulink®opens a new empty simulation file (Fig. 2.3). First, the file must be saved. Saving the file can be done with the aid of **File** > **Save**, as shown in Fig. 2.4. We saved the file as "Buck.slx" since we want to simulate a Buck converter as our first example.

Figure 2.1: "simulink" command opens the Simulink®environment.

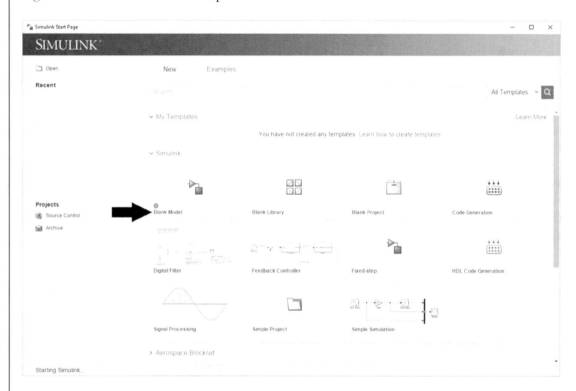

Figure 2.2: Simulink's first window.

The "Library Browser" icon (Fig. 2.5) is one of the most important icons in the Simulink®environment. With the aid of the "Library Browser" icon, one can place the required blocks to the simulation file.

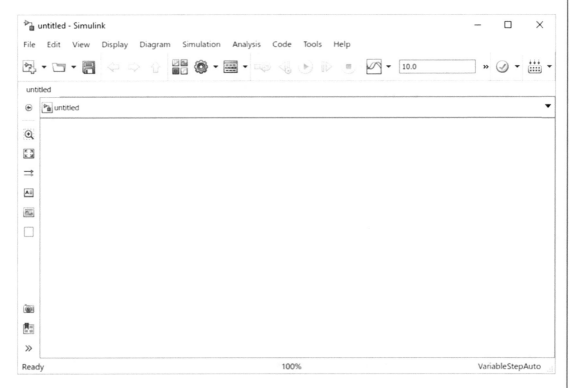

Figure 2.3: Simulink®environment.

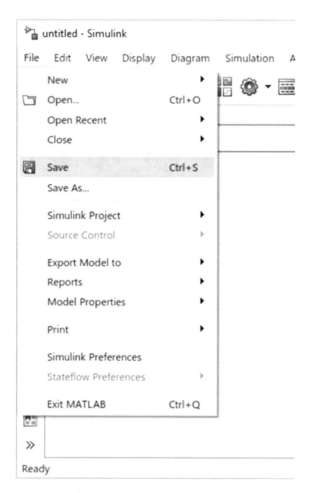

Figure 2.4: File>Save is used to save the simulation file.

Figure 2.5: "Library Browser" icon.

After clicking the "Library Browser" icon, a window containing the blocks will appear on the screen (Fig. 2.6).

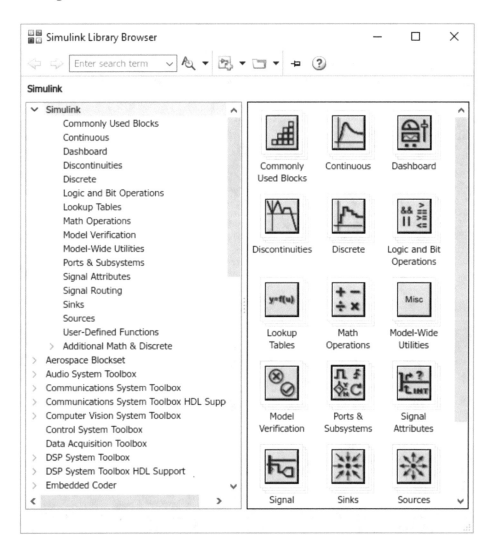

Figure 2.6: Simulink®libraries.

You can use the "Enter search term" box to search for a block (Fig. 2.7). For example, if you need an integrator block but forgot its place, you can simply write "integrator" and press Enter (Fig. 2.8). Simulink®, search for the entered term and the obtained results on the right of window are shown.

Figure 2.7: "Enter search term" box.

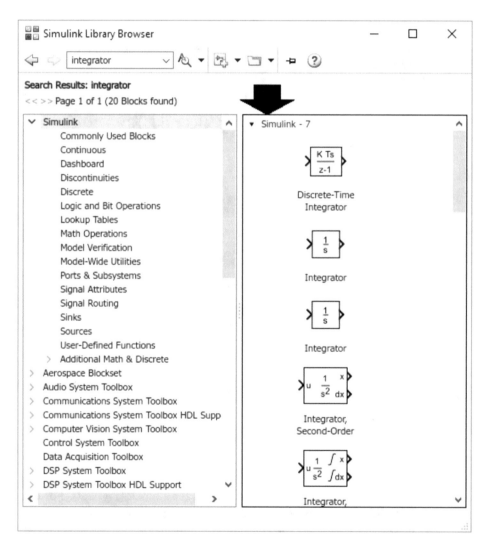

Figure 2.8: Search result for "integrator" block.

Simulation of DC-DC and other types of power electronics converters can be done with the aid of the Simscape library (Fig. 2.9). Simscape can simulate systems containing electronic, termal, mechanical, etc. parts. We use the **Power System** > **Specialized Technology** tab to simulate the power electronics converters (Fig. 2.9).

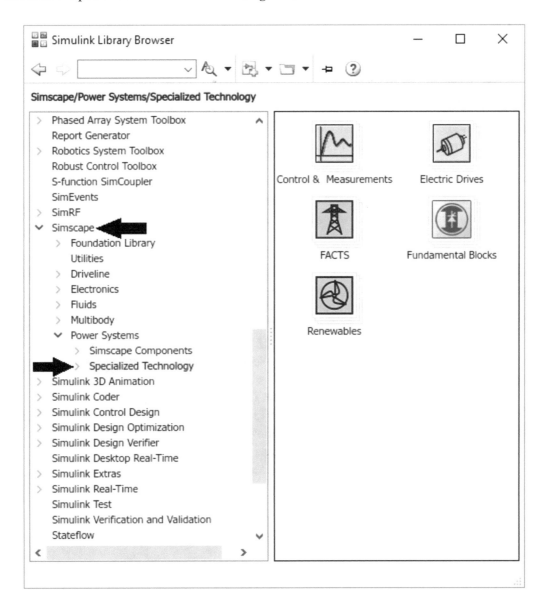

Figure 2.9: "Specialized Technology" block set.

As can be seen from Fig. 2.10, the "Specialized Technology" tab contains five sub-categories:

- Fundamental Blocks

- Control and Measurements

- Electric Drives

- FACTS

- Renewables

From the DC-DC converter simulation viewpoint the most important are: Control and Measurements and Fundamental Blocks. As the name suggests, "Control and Measurements" (see Fig. 2.10) contains blocks like: THD, RMS, average, Fourier, and Pulse Width Modulation (PWM).

Elements such as resistors, capacitors, inductors, sources, different types of switches like MOSFET's, diodes, tristors, etc., can be found in "Fundamental Blocks" (Fig. 2.11) menu.

Here is a quick review of blocks available in the "Fundamental Blocks" section.

If we click on the "Electrical Sources" icon, the window shown in Fig. 2.12 will open. Figure 2.12 shows the different type of sources available in the "Electrical Sources" section of "Fundamental Blocks." Returning to the "Fundamental Blocks" menu can be done either by clicking on the small arrow shown in Fig. 2.13 or by clicking the "Fundamental Blocks" (Fig. 2.14).

After clicking the "Elements" icon (Fig. 2.11), the window shown in Fig. 2.15 is opened. Different types of loads can be found here. The most useful one is "Series RLC Branch" (Fig. 2.16). Elements such as: R, L, C, series RL and series RC elements can be simulated with the aid of the "Series RLC Branch" block. We use this block to simulate the passive components in the converter.

If we click on the "Power Electronics" icon (Fig. 2.11), the window shown in Fig. 2.17 will open. Switches such as MOSFET, Diode, BJT, IGBT, etc. can be found here. If we click on the "Measurement" icon (Fig. 2.11), the window shown in Fig. 2.18 will open. Blocks like the amper meter and voltmeter can be found here.

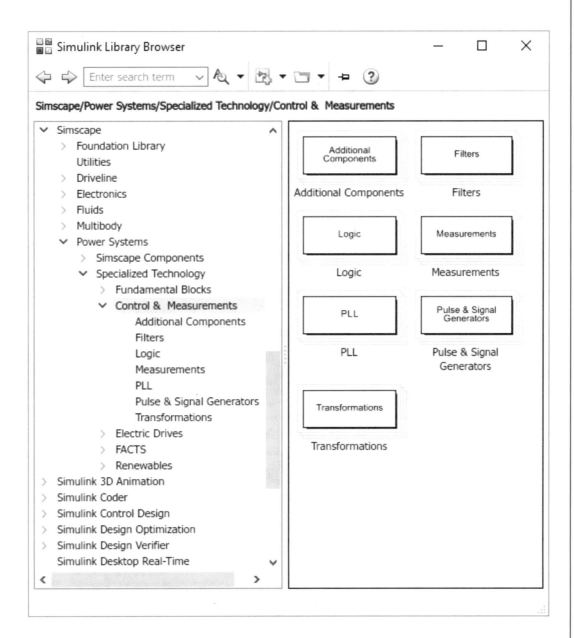

Figure 2.10: "Control & Measurements" block set.

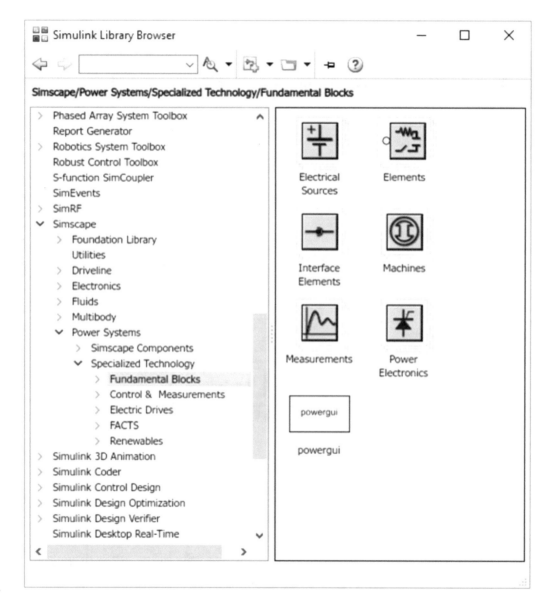

Figure 2.11: "Fundamental Blocks" block set.

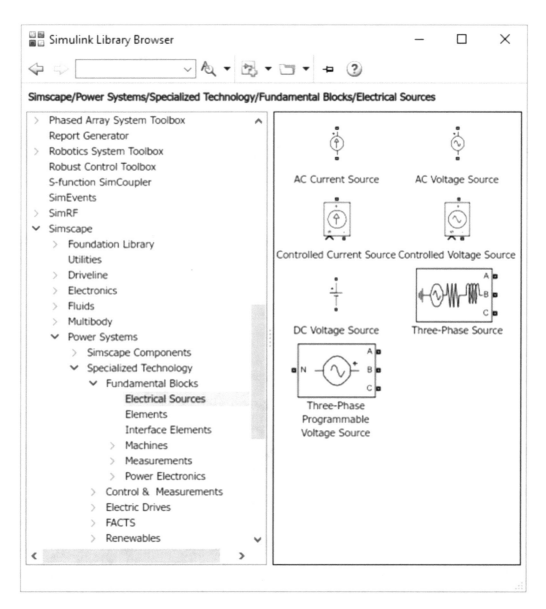

Figure 2.12: "Electrical Sources" block set.

Figure 2.13: Return icon.

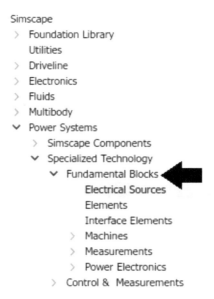

Figure 2.14: You can return to previous windows by clicking the "Fundamental Blocks," as well.

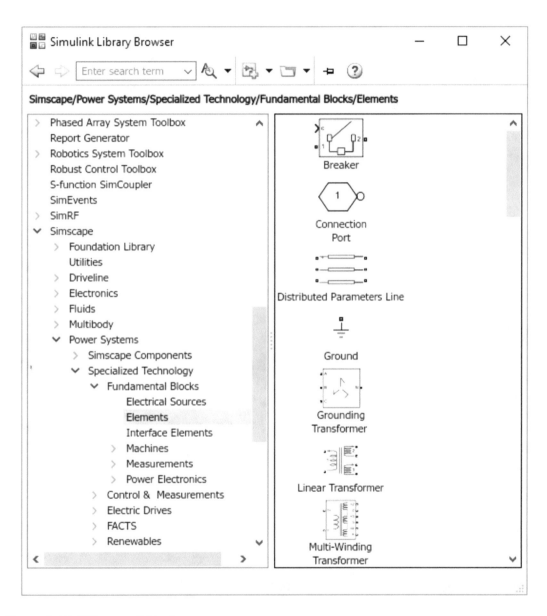

Figure 2.15: "Elements" block set.

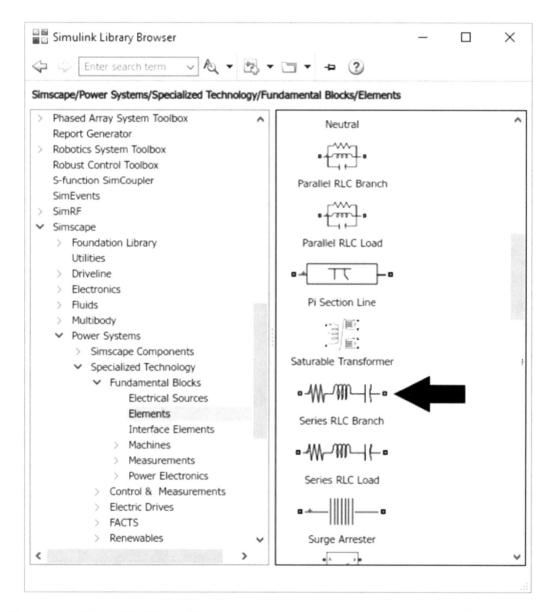

Figure 2.16: "Series RLC Branch" is used to simulate R, L, C, RL, RC, LC and open circuit.

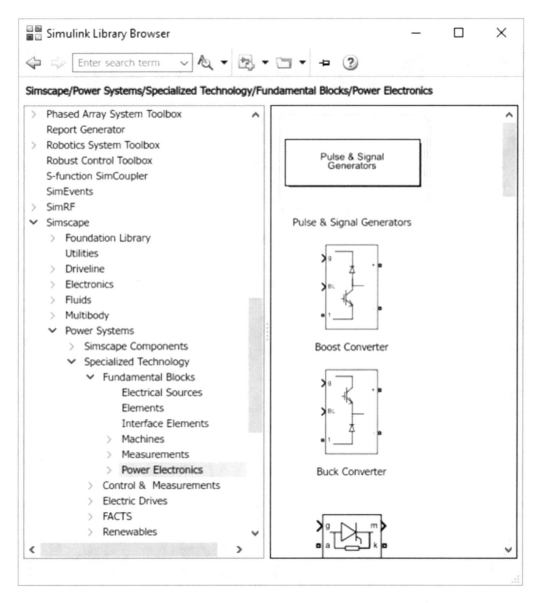

Figure 2.17: "Power Electronics" block set.

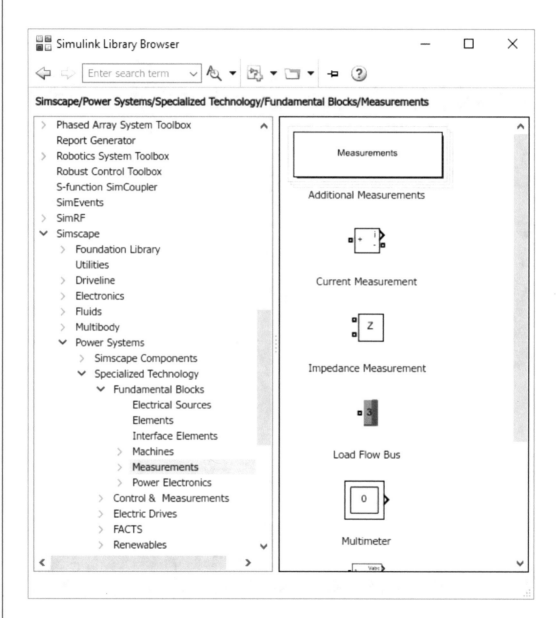

Figure 2.18: "Measurements" block set.

2.3   SIMULATION OF A BUCK CONVERTER

In this section we want to simulate a Buck converter shown in Fig. 2.19 with the aid of Simulink®.

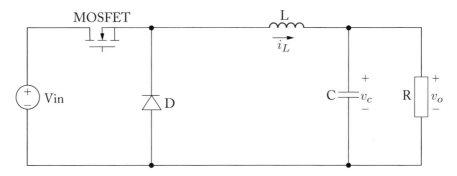

Figure 2.19: Buck converter circuit.

Here are the elements values: $V_{in} = 50$ V, $r_{ds} = 0.1$ Ω, $r_D = 0.01$ Ω, $V_D = 0.7$ V, $L = 400\,\mu$H, $C = 100\,\mu$F, and $R = 5$ Ω. r_{ds}, r_D, and V_D shows MOSFET on resistance (resistance between drain and source), forward-biased diode resistance, and forward-biased diode voltage drop, respectively. A forward-biased diode is modeled as shown in Fig. 2.20.

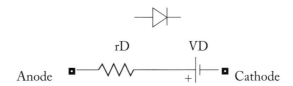

Figure 2.20: Linear model for diode.

In order to simulate the Buck converter, we add the "DC Voltage Source" (Fig. 2.12) to the working area by drag and drop. As soon as we drop the "DC Voltage Source," Simulink®asks the amplitude of voltage source (Fig. 2.21). We can enter the value we need (in this example 50) or we can postpone it to later. We click on an empty point of working area so Simulink®assigns the predefined value of 100 V to the source. We change this value later.

Rotation of a placed element can be done with the aid of **Ctrl+R**. First, the element is selected by clicking on it. After clicking, a blue rectangle is drawn around the element which shows that the element is selected (Fig. 2.22).

Pressing the Ctrl and R keys simultaneously cause the element to rotate (Fig. 2.23). You can rotate an element 180° by selecting it (Fig. 2.24) and pressing **Ctrl+I**.

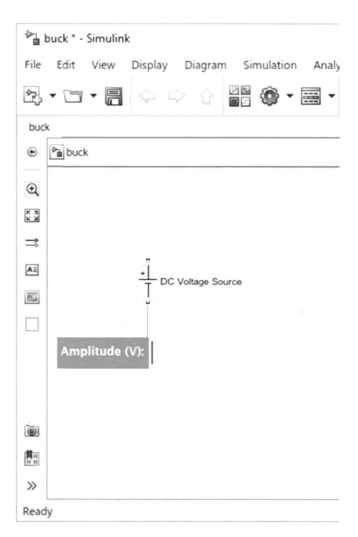

Figure 2.21: Dropping the "DC Voltage Source" into the working area.

Figure 2.22: Vertical "DC Voltage Source" block.

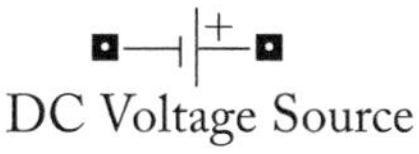

Figure 2.23: Result of pressing Ctrl+R in Fig. 2.22.

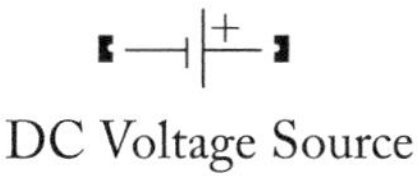

Figure 2.24: Horizontal "DC Voltage Source" block.

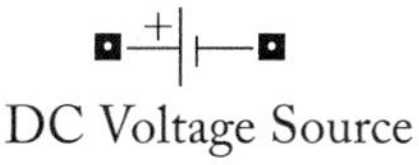

Figure 2.25: Result of pressing Ctrl+I in Fig. 2.24.

There is another way to rotate an element. First, we select the element by left clicking on it. After selection is done, right click on it and select **Rotate & Flip**. A menu containing rotation options will appear (Fig. 2.26), allowing you to select the desired choice from the appeared list.

You can change the element name by clicking on the default name (DC Voltage Source) and writing the new name. As shown in Fig. 2.27, the voltage source name is changed to "VIN."

If you do not want to show the elements name, just left click on it. After selection (blue rectangle is appeared), right click on it and select the **Format** and uncheck the "Show Block Name." As shown in Fig. 2.29, the block name no longer appears.

If you need help about an element you can double click on the element and press the "Help" button in the opened dialog (Fig. 2.30). This opens the document about that specific element. We add the rest of required parts to the simulation file (Fig. 2.31).

We change the elements default names and connect the elements together, as shown in Fig. 2.32. When the mouse pointer is placed on the either of elements ends, the pointer is changed to +. This shows you can start connecting that end to other parts. To connect two elements together, the left mouse button is pressed at the source and is released at the destination.

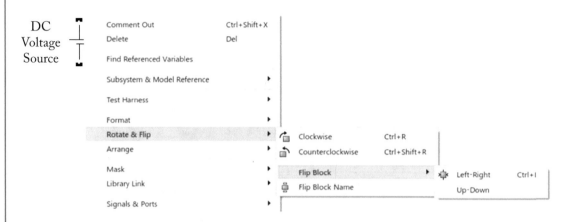

Figure 2.26: "Rotate & Flip" menu.

Figure 2.27: Changing the name of the block.

Figure 2.28: Uncheck the "Show Block Name" option.

Figure 2.29: Simulink®will not show block names after you uncheck the "Show Block Name" option.

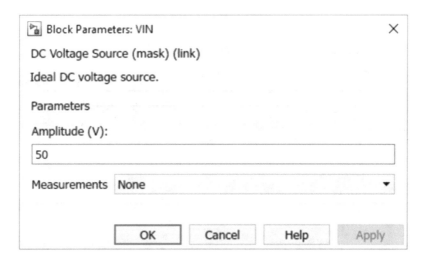

Figure 2.30: "VIN" block settings.

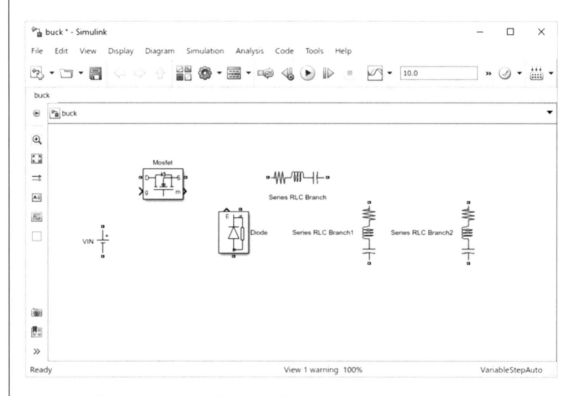

Figure 2.31: Placing the required blocks into the working area.

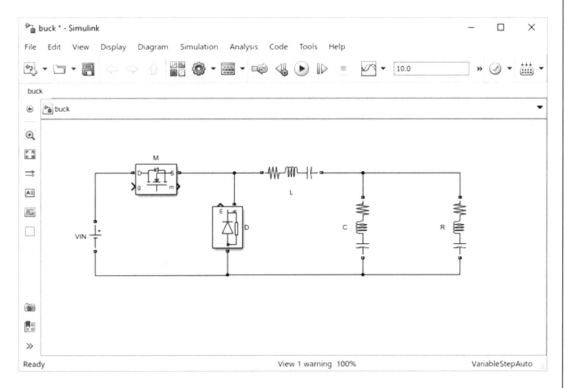

Figure 2.32: Connecting the elements together.

After connecting the elements together, Series RLC Branches must be converted into suitable elements. For example, a Series RLC Branch named "L" (Fig. 2.32) must be converted into an inductor. To do this, we double click on the Series RLC Branch named "L" (Fig. 2.33) and change the "Branch type" to "L", as shown in Fig. 2.34.

We use the same procedure to convert other Series RLC Branches into suitable elements. Figure 2.35 shows the final result. The values of elements are set by double clicking the element and entering the desired values into the boxes. For example, the inductor value is set to 400 μF by entering 400e-6 in the Inductance (H) box (Fig. 2.36). 400e-6 in MATLAB environment equals to 400×10^{-6}. If you like to set the inductor initial current, "Set the initial inductor current" must be checked (Fig. 2.37). The same procedure can be used to set the capacitor initial voltage. Other elements values are set in the same way.

MOSFET on resistance (r_{ds}) is set by the "FET resistance Ron (Ohms):" box. As shown in Fig. 2.20, the forward-biased diode is modeled with a resistor in series with a voltage source. Value of series resistor (rD) and voltage drop (VD) is set by "Resistance Ron (Ohms):" and "Forward voltage Vf (V):" boxes, respectively (Fig. 2.39). "Resistance Ron (Ohms):" and "Forward voltage Vf (V):" boxes sets to 0.01 Ω and 0.7 V, respectively.

Output voltage is measured with the aid of a "Voltage Measurement" block (Fig. 2.40). The "Voltage Measurement" block acts as a voltmeter. We connect it to the loads terminals to measure the load voltage (Fig. 2.41). Output is observed with the aid of a "Scope" block (Fig. 2.42). Output of "Voltage Measurement" block is connected to the scope (Fig. 2.43).

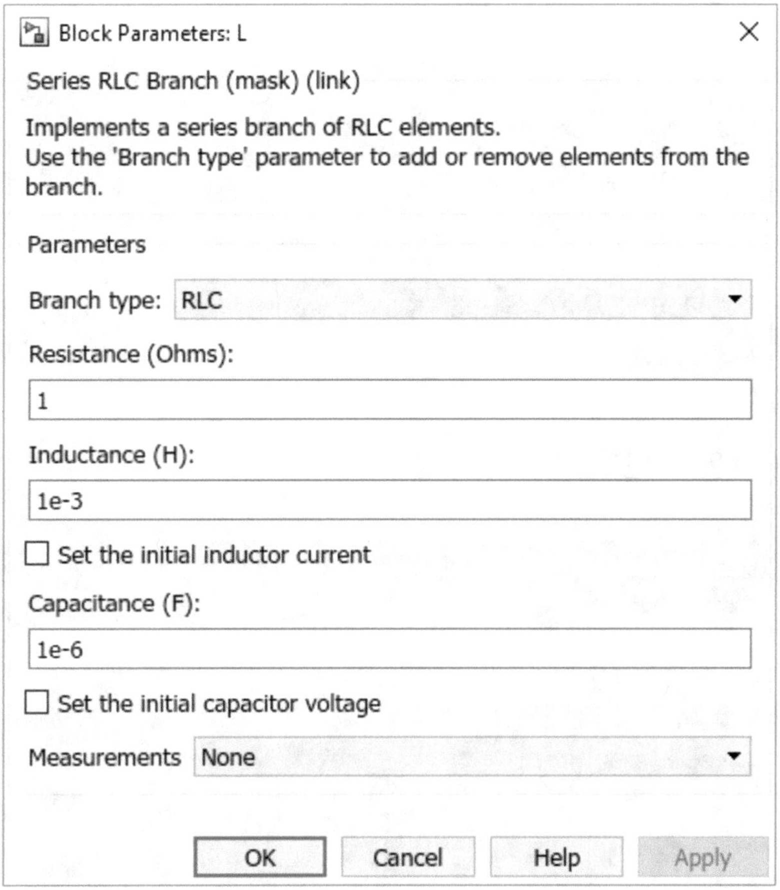

Block Parameters: L   ×

Series RLC Branch (mask) (link)

Implements a series branch of RLC elements.
Use the 'Branch type' parameter to add or remove elements from the
branch.

Parameters

Branch type: RLC   ▼

Resistance (Ohms):

1

Inductance (H):

1e-3

☐ Set the initial inductor current

Capacitance (F):

1e-6

☐ Set the initial capacitor voltage

Measurements None   ▼

OK   Cancel   Help   Apply

Figure 2.33: "Series RLC Branch" settings.

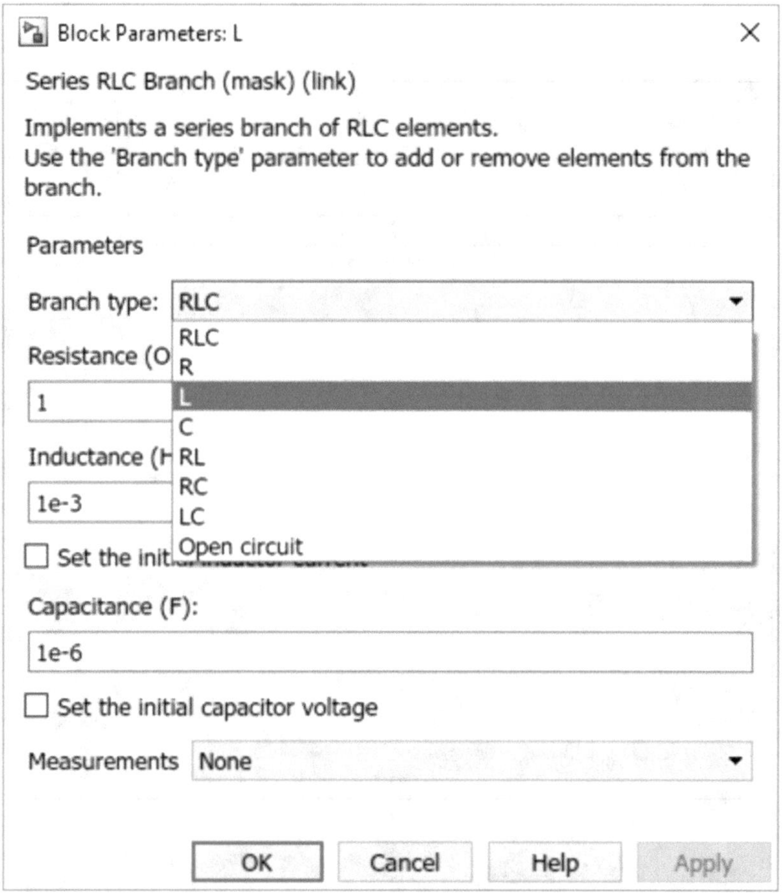

Figure 2.34: Conversion of "Series RLC Branch" into an inductor.

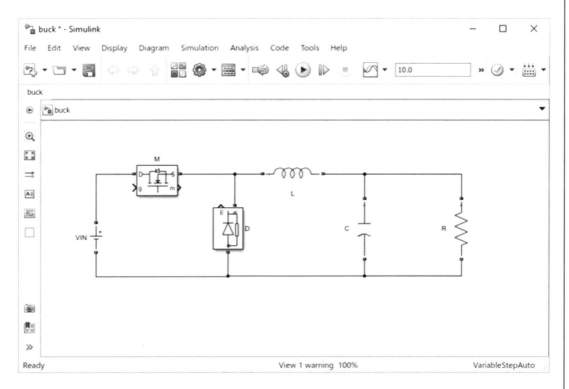

Figure 2.35: Final Buck converter in Simulink®.

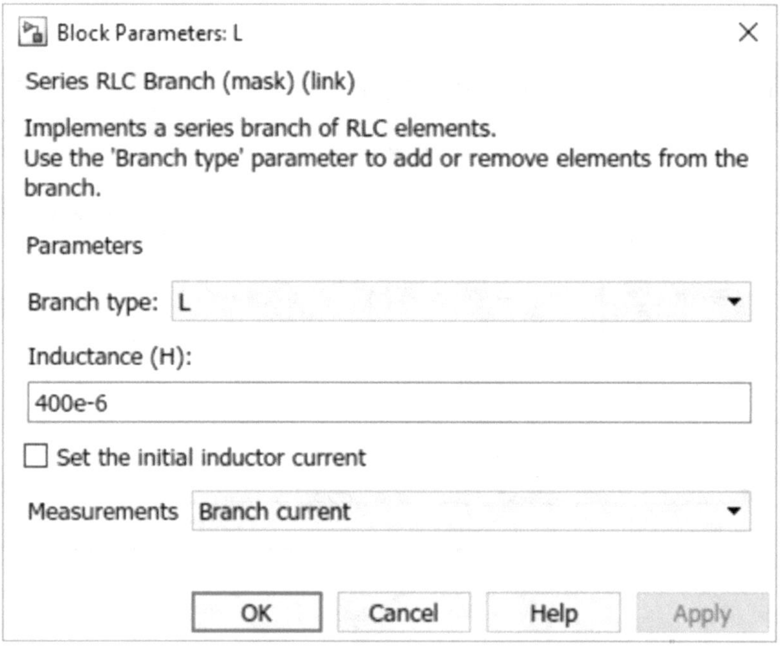

Figure 2.36: Setting the inductance to 400 μF is done by entering 400e-6 in the "Inductance (H)" box.

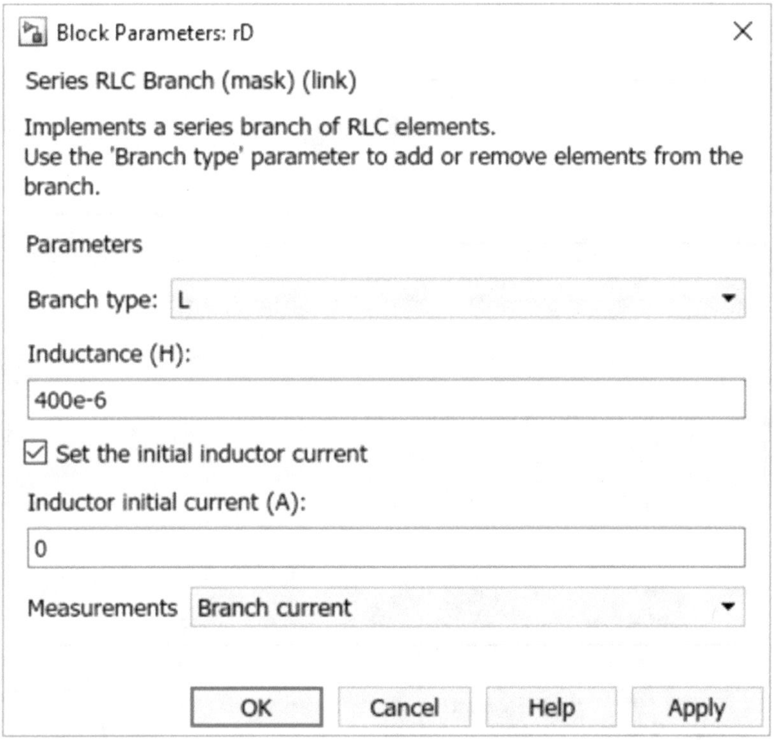

Figure 2.37: Setting the initial current.

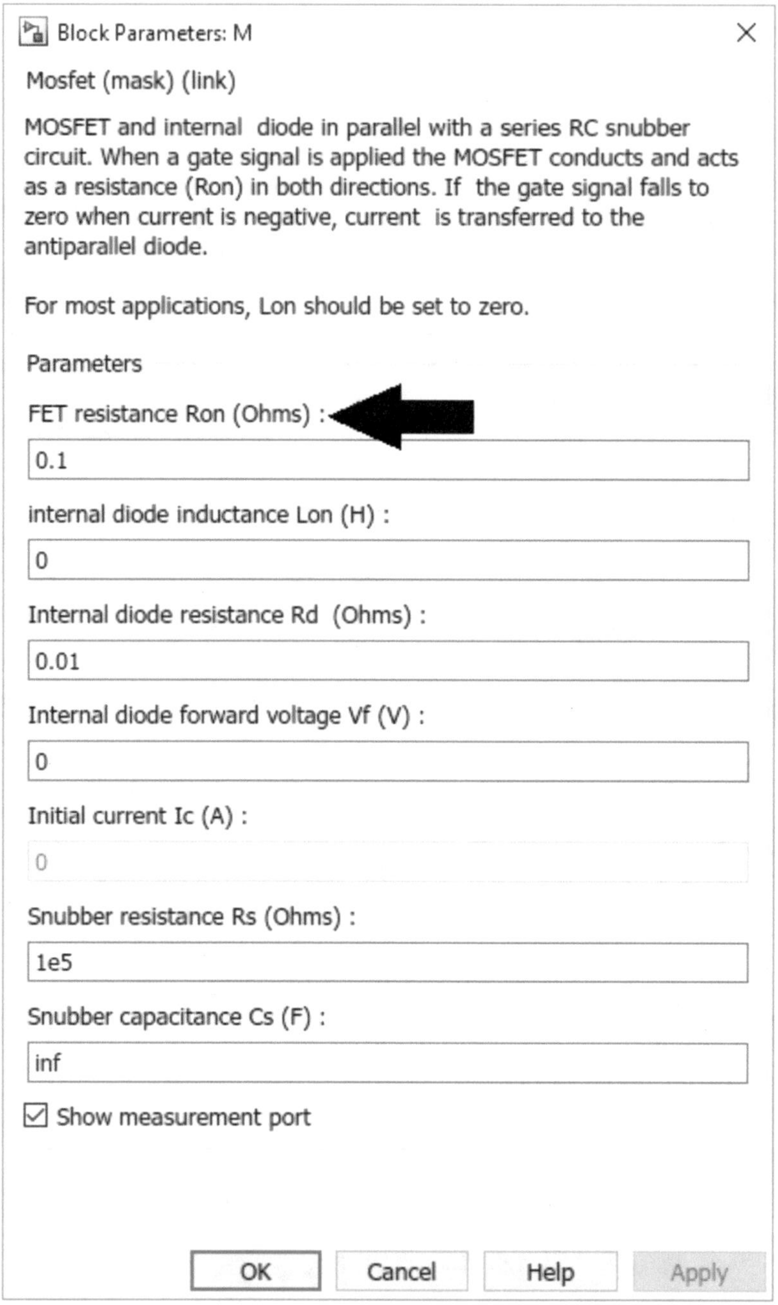

Figure 2.38: "MOSFET" settings.

Block Parameters: D   ✕

Diode (mask) (link)

Implements a diode in parallel with a series RC snubber circuit.
In on-state the Diode model has an internal resistance (Ron) and
inductance (Lon).
For most applications the internal inductance should be set to zero.
The Diode impedance is infinite in off-state mode.

Parameters

Resistance Ron (Ohms) :

| 0.01 |

Inductance Lon (H) :

| 0 |

Forward voltage Vf (V) :

| 0.7 |

Initial current Ic (A) :

| 0 |

Snubber resistance Rs (Ohms) :

| 500 |

Snubber capacitance Cs (F) :

| 250e-9 |

☑ Show measurement port

| OK | Cancel | Help | Apply |

Figure 2.39: Diode settings.

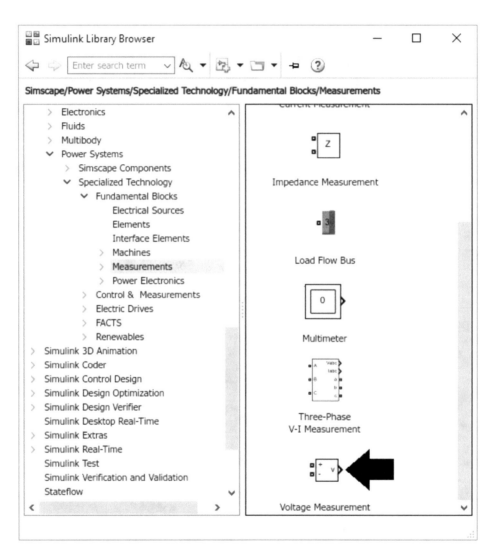

Figure 2.40: Simulink's volt meter block.

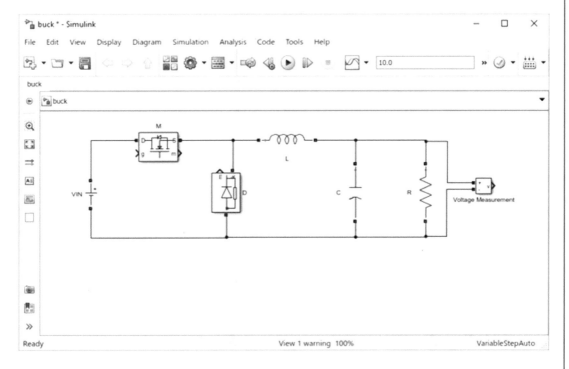

Figure 2.41: Adding the "Voltage Measurements" block to the simulation file.

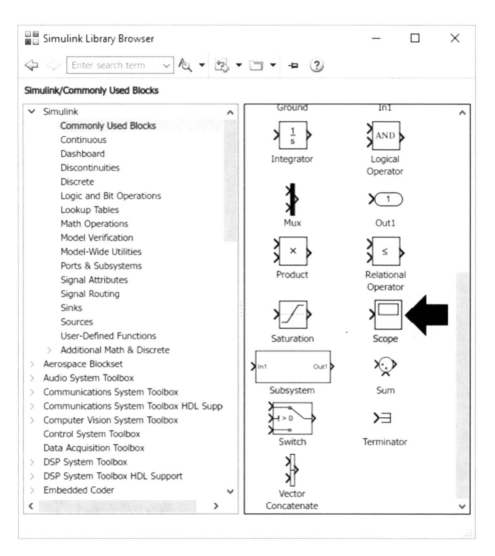

Figure 2.42: "Scope" block.

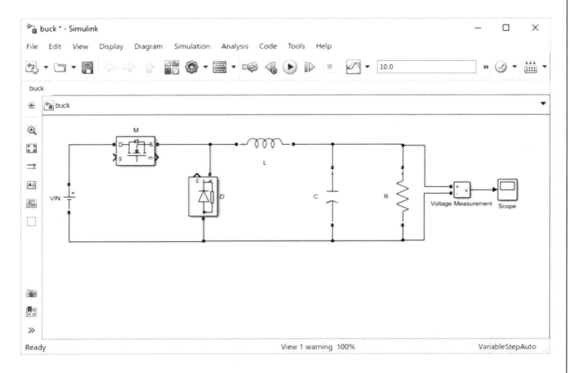

Figure 2.43: Buck converter power stage without MOSFET gate signals.

Up until now, the skeleton of the Buck converter is ready. The next step is to provide the required gate signal for MOSFET. This can be done in a number of different ways. The simplest is to use the "Pulse Generator" block, as shown in Fig. 2.44. We connect the output of "Pulse Generator" block to the "g" terminal of MOSFET, as shown in Fig. 2.45. Assume that we want to simulate the circuit with a duty ratio of 41% and switching frequency is taken as 50 KHz. With this assumption in mind, "Pulse Generator" s, "Period (sec):" and "Pulse Width (% of period)" boxes are filled, as shown in Fig. 2.46.

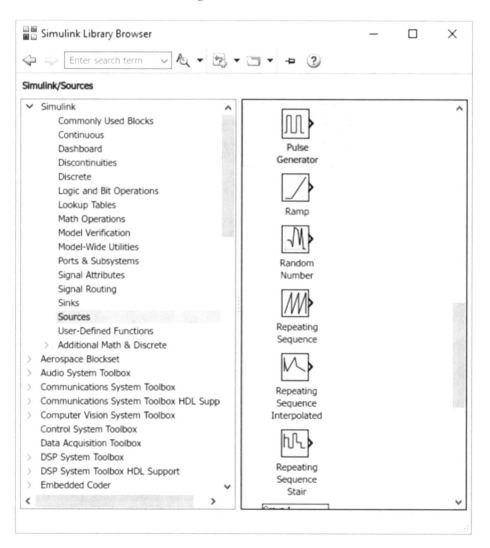

Figure 2.44: "Pulse Generator" block.

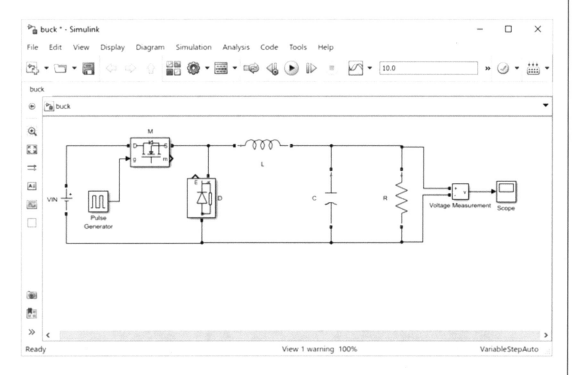

Figure 2.45: Finalized Buck converter simulation file with "Pulse Generator" block.

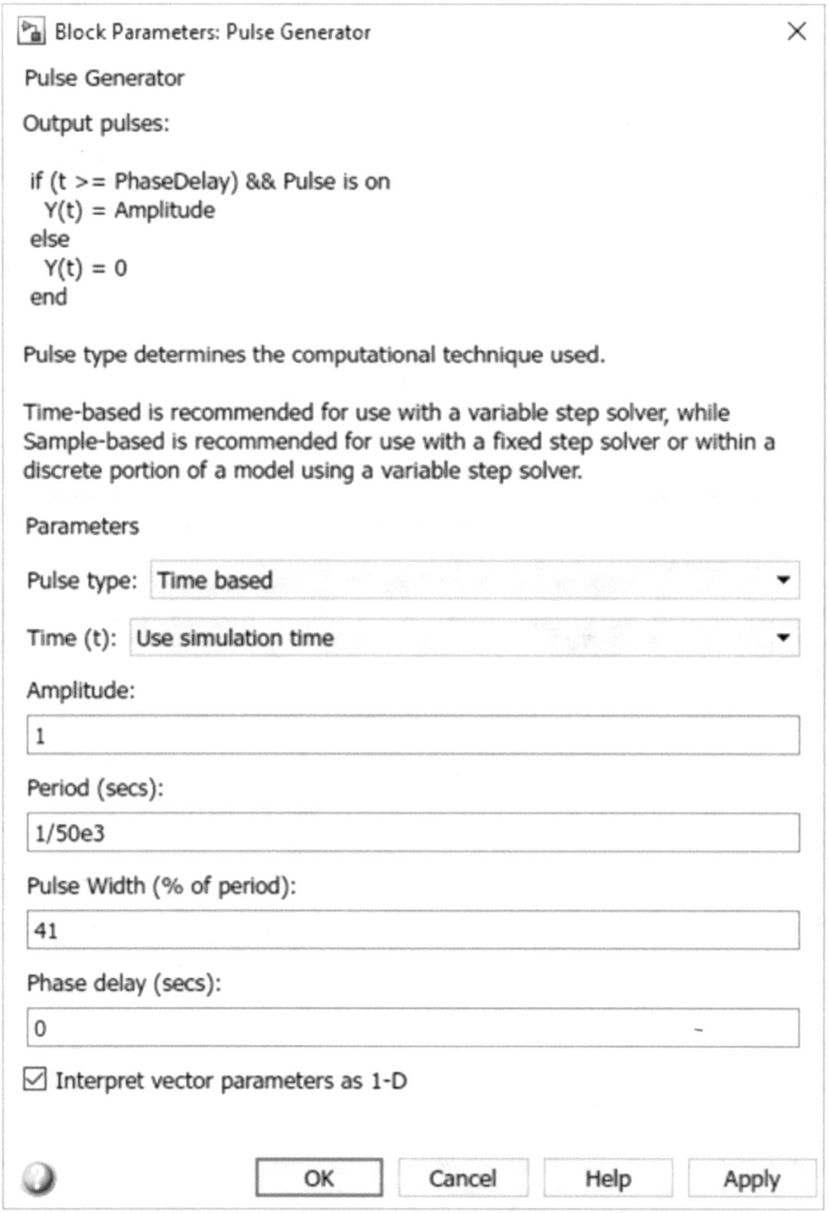

Block Parameters: Pulse Generator   ✕

Pulse Generator

Output pulses:

if (t >= PhaseDelay) && Pulse is on
  Y(t) = Amplitude
else
  Y(t) = 0
end

Pulse type determines the computational technique used.

Time-based is recommended for use with a variable step solver, while
Sample-based is recommended for use with a fixed step solver or within a
discrete portion of a model using a variable step solver.

Parameters

Pulse type: Time based   ▼

Time (t): Use simulation time   ▼

Amplitude:

1

Period (secs):

1/50e3

Pulse Width (% of period):

41

Phase delay (secs):

0   ˷

☑ Interpret vector parameters as 1-D

🌑   OK   Cancel   Help   Apply

Figure 2.46: "Pulse Generator" settings.

There is a ready-to-use PWM generator in Simulink®(Fig. 2.47). After setting the switching frequency, all that must be done is to tell it the duty ratio you need. If you prefer to use the block you can add the "PWM Generator (DC-DC)" to your simulation instead of "Pulse Generator" block, as shown in Fig. 2.48. Double clicking the "PWM Generator (DC-DC)" block open the dialog shown in Fig. 2.49. Here, you can enter the desired switching frequency (in this example 50 KHz).

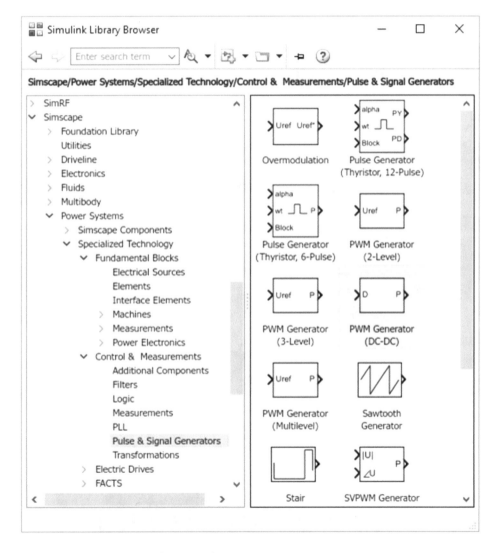

Figure 2.47: PWM Generator (DC-DC) block.

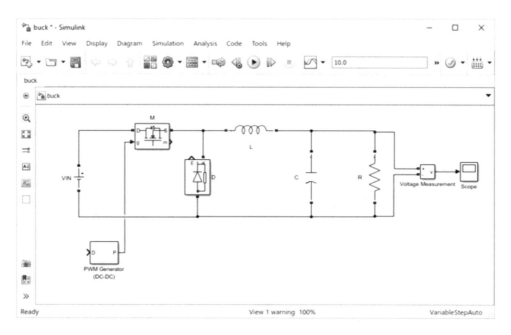

Figure 2.48: Adding the "PWM Generator (DC-DC)" to the simulation file.

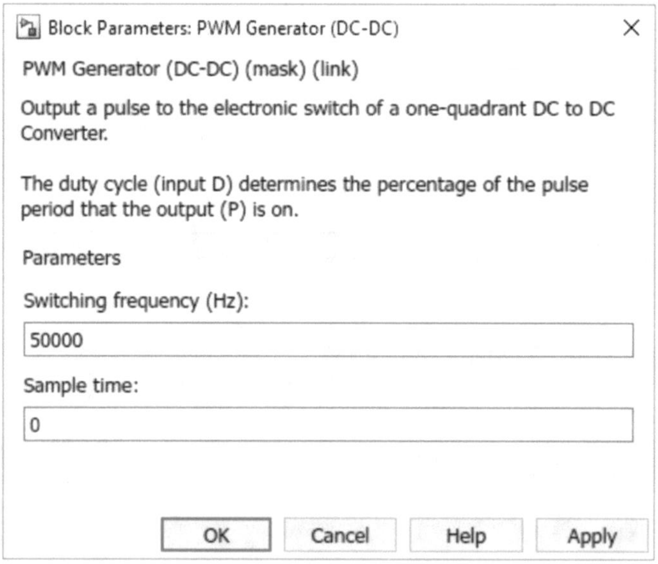

Figure 2.49: "PWM Generator (DC-DC)" block settings.

Desired duty ratio (in this example 0.41) is set by a constant block (Fig. 2.50). We connect a "Constant" block with value of 0.41 to the input of "PWM Generator (DC-DC)" block (Fig. 2.51). Input of "PWM Generator (DC-DC)" block must always be greater than zero and less than one. To satisfy this condition you can place a "Saturation " block before it (Fig. 2.52). "Saturation" block can be found in the "Discontinuities" section as shown in Fig. 2.53. If you want to place the "Saturation" block before the "PWM Generator (DC-DC)" block like that shown in Fig. 2.52, you must do the "Saturation" block settings like that shown in Fig. 2.54.

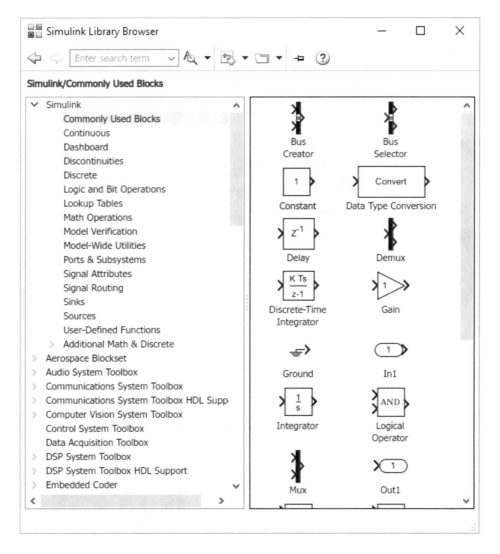

Figure 2.50: "Constant" block.

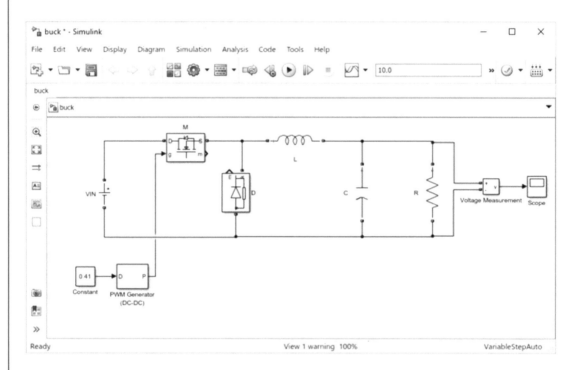

Figure 2.51: Finalized Buck converter with "PWM Generator (DC-DC)" block.

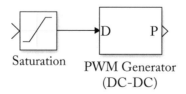

Figure 2.52: Adding the "Saturation" block to the "PWM Generator (DC-DC)" block.

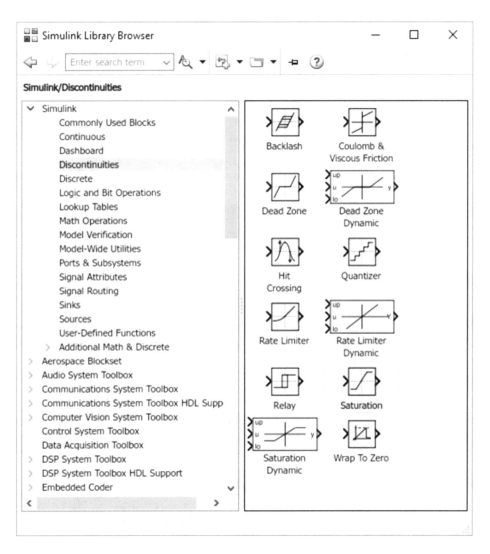

Figure 2.53: "Saturation" block.

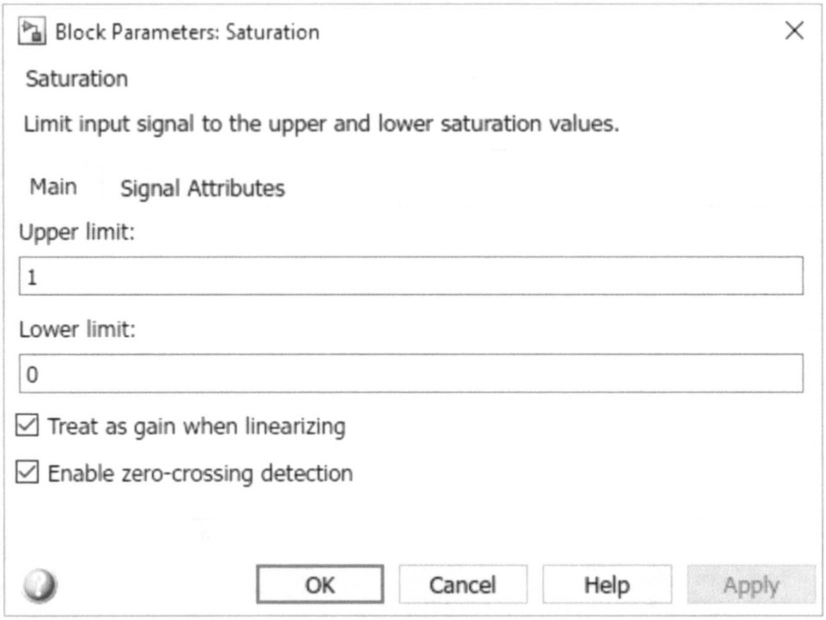

Figure 2.54: "Saturation" block settings.

Assume that we want to simulate the circuit for 10 ms. Desired simulation length is set by the box shown in Fig. 2.55.

Figure 2.55: Setting the desired simulation length.

The simulation diagram seems complete. The last step is running the simulation. Running the simulation is done by clicking the button shown in Fig. 2.56. If we click the button an error message is shown (Fig. 2.57).

The reason we face such a windows is that the "powergui" block (Fig. 2.58) had not been added to the simulation. We add the "powergui" block to the simulation and re-run the simulation. Simulation is done without any error. If we double click the scope block, we see the load voltage.

Figure 2.56: Running the simulation.

Figure 2.57: An error message is produced when there is no "powergui" block in the simulation file.

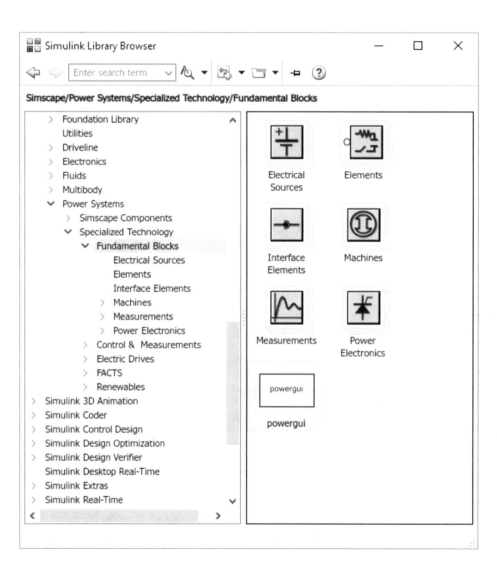

Figure 2.58: "powergui" block.

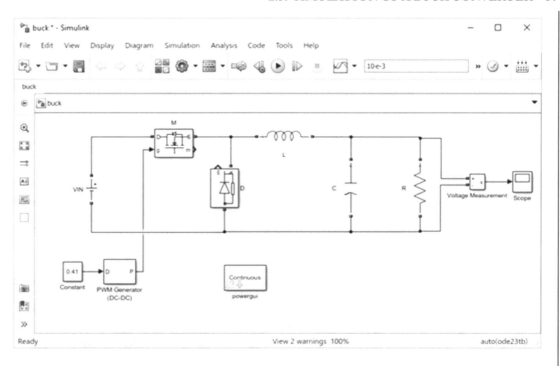

Figure 2.59: Ready to use simulation file.

As shown in Fig. 2.60, the background of scope is black and the curve is drawn in yellow. You can change theses colors if you like. First, the small arrow behind the gear wheel is clicked (Fig. 2.61). We select the "Style" from the appeared menu. This bring the color selection dialog box. For example, if we select the colors shown in Fig. 2.62, the appearance shown in Fig. 2.63 is obtained. All the scope graphs in this book are produced with the setting shown in Fig. 2.62. So, all the book graphics have the white background.

If you double click the "powergui" block, you will see a dialog box like that shown in Fig. 2.64.

"Simulation type:" can be set to "Continuous," "Discrete," or "Phasor." Here you tell Simulink®how the equation behind the circuit must be solved. Generally, "Continuous" or "Discrete" is used for converter simulation. If you select the "Continuous" option, methods like Runge-Kutta is used to solve the circuit differential equation. You can select your desired method but it is better to use "stiff" solver. In order to choose a solver, click on the gear wheel shown in Fig. 2.65.

You can select your desired method from the list of available solvers as shown in Figs. 2.66 and 2.67. "ode23s" or "ode23t" is enough for a large class of simulations.

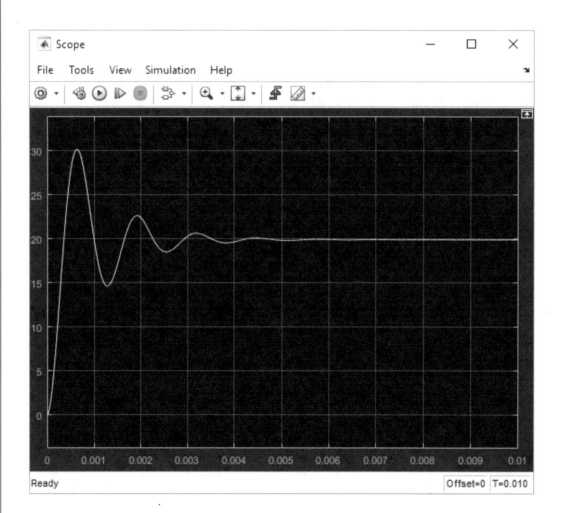

Figure 2.60: Simulation result.

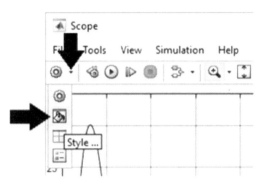

Figure 2.61: Changing the "Scope" blocks settings.

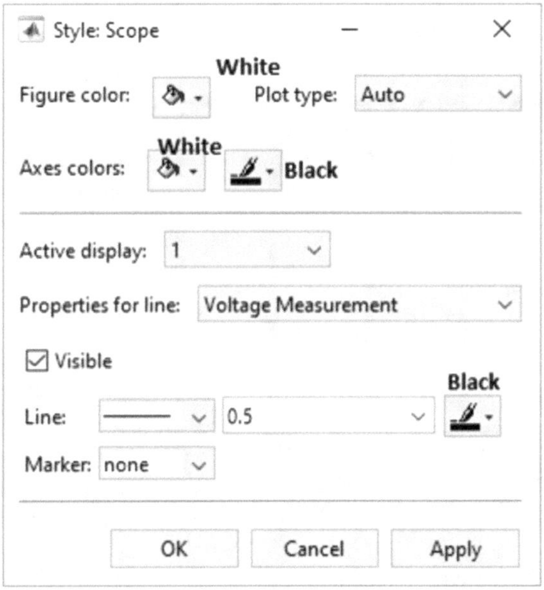

Figure 2.62: Scope block colors.

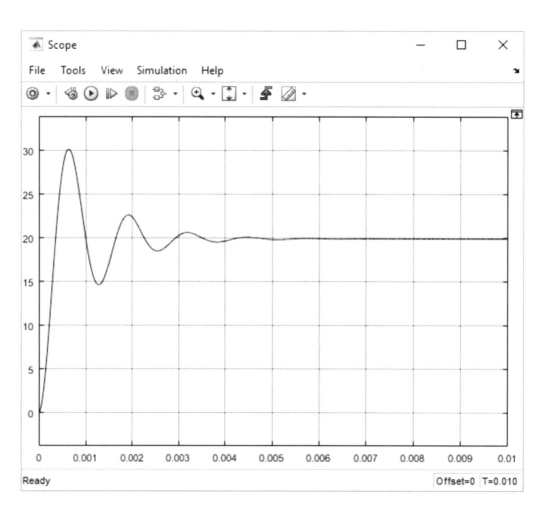

Figure 2.63: Scope block with white background.

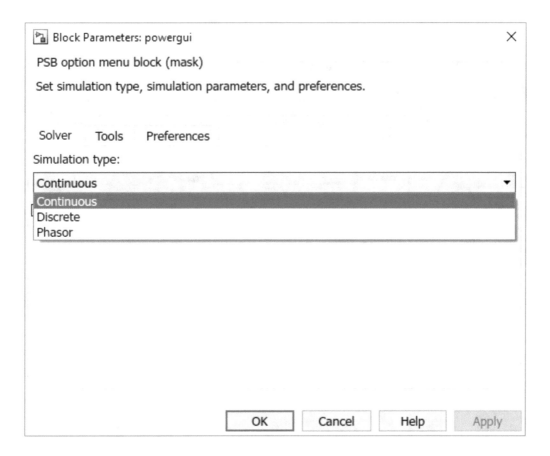

Figure 2.64: "powergui" block settings.

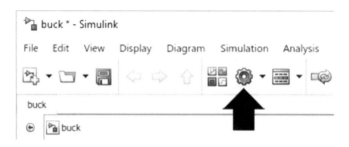

Figure 2.65: "Model Configuration Parameters" icon.

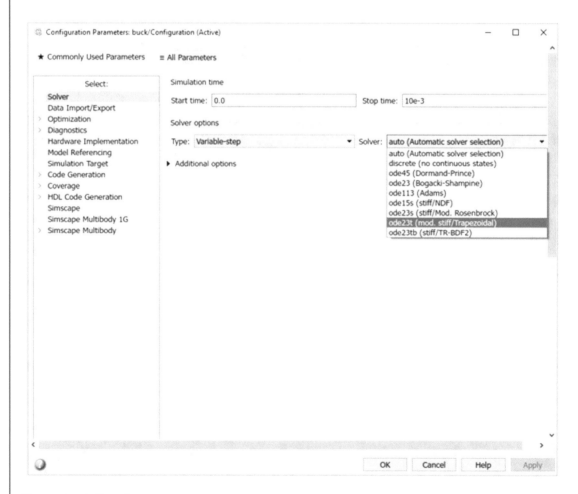

Figure 2.66: Simulation settings.

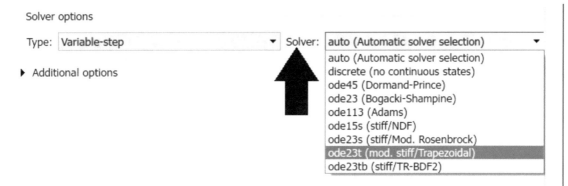

Figure 2.67: Available "Solver" list of Simulink®.

If you select the "Discrete" option (Fig. 2.64), the differential equation is discretized and the discrete version is solved. If you want to use the "Discrete" option, you must choose the "Sample time (s)" small enough. "Sample time (s)" define the time step used in solving the resultant discrete equation. For example, if the switching frequency is f_s, "Sample time (s)" of $\frac{1}{N \times f_s}$ where $N \geq 20$ seems enough. If you take a very small time step, obtained waveforms are more accurate but you must wait a longer time to complete the simulation since number of required calculation increase.

"Phasor" (Fig. 2.64) is not a suitable for DC-DC converter simulation. It is useful for power system simulations.

If you want to observe the inductor current you can add an amper meter (Fig. 2.69) to the circuit, as shown in Fig. 2.70. The inductor current is shown in Fig. 2.71. As you see, converter works in Continuous Current Mode (CCM).

There is another method to measure an element current. This method does not to add amper meter block. We double click on the element and select the desired voltage/current from the "Measurements" drop-down list. For instance if you want to observe the current in the inductor, you double click on the inductor and select the "Branch current" as shown in Fig. 2.72. You must add a "Multimeter" block (Fig. 2.73) to the simulation file as shown in Fig. 2.74. "Multimeter" block can catch the inductor current so you can see it with the aid of a scope. In order to ask the "Multimeter" block to gather the inductor current information, you must double click on the "Multimeter" block (Fig. 2.75). You click on the "Ib: L" from the left list, then you click on ">>" to add it to the right list. The last step is clicking the "Close" button (Fig. 2.76). As you may notice, the elements have a red + in one of the terminals (Fig. 2.77). When current enters the terminal with red +, the current is assumed to be positive. The grey arrow in Fig. 2.77 shows the positive inductor current.

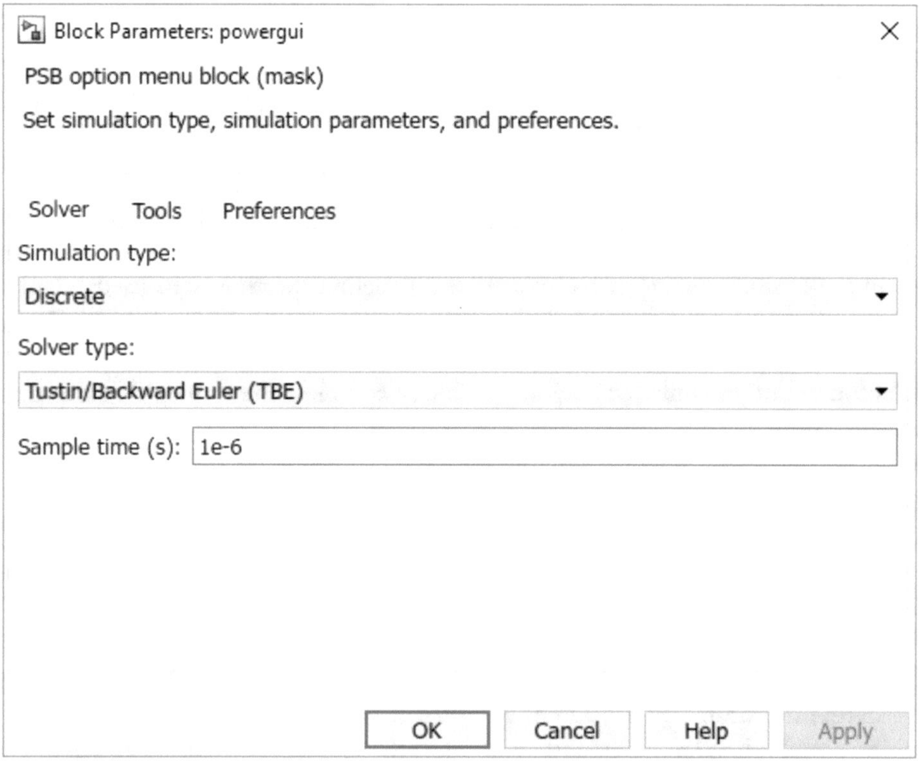

Figure 2.68: "Discrete" simulation type.

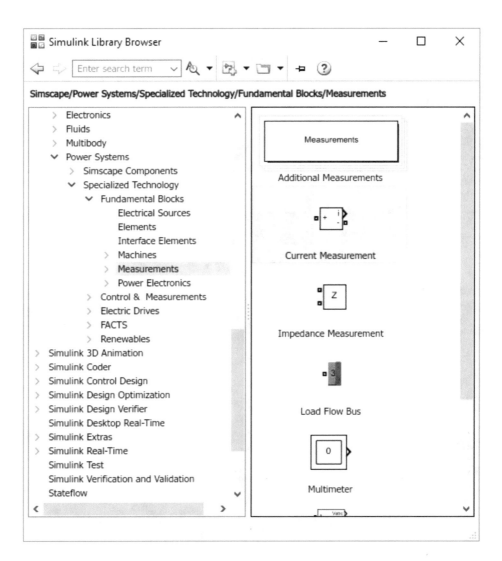

Figure 2.69: **Amper meter block.**

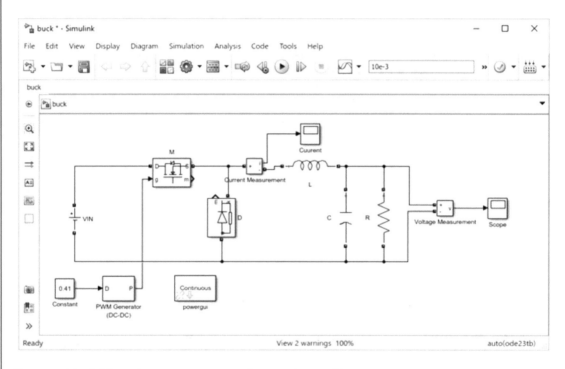

Figure 2.70: Adding the amper meter to the simulation file.

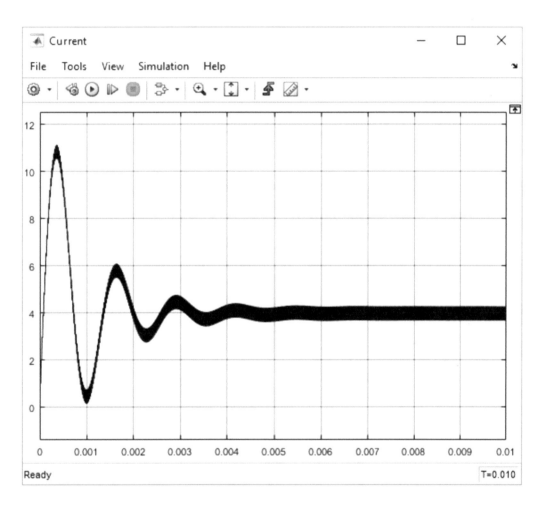

Figure 2.71: Inductor current.

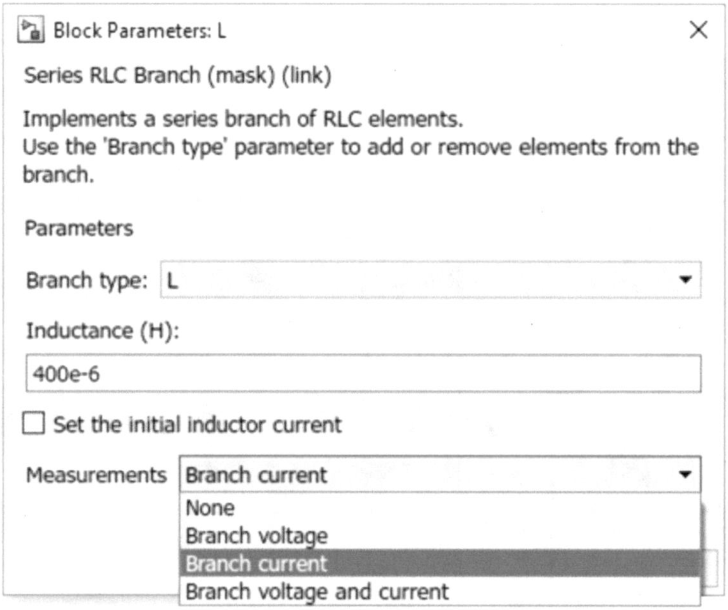

Figure 2.72: Using the "Measurements" drop down list to measure the inductor current.

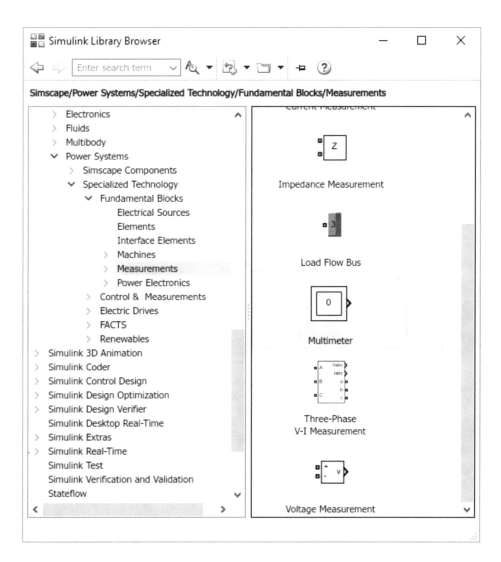

Figure 2.73: "Multimeter" block.

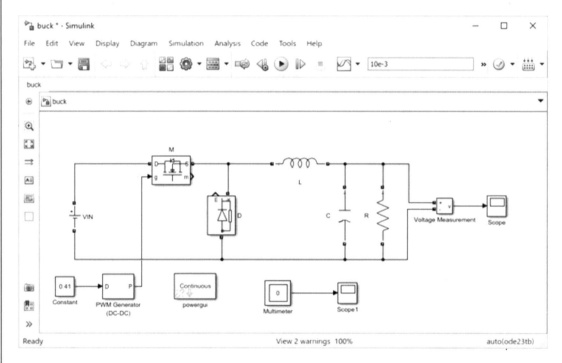

Figure 2.74: Adding the "Multimeter" block to the simulation file.

Figure 2.75: "Multimeter" block setting.

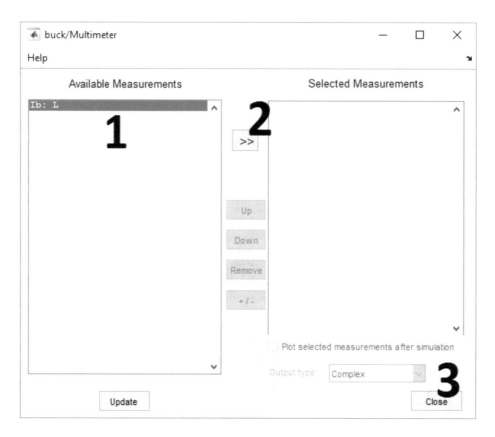

Figure 2.76: Adding the desired signals to the "Multimeter" block.

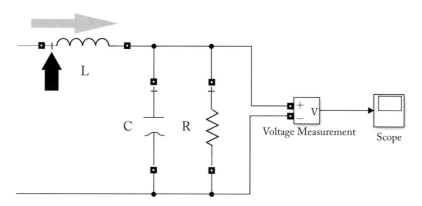

Figure 2.77: Positive direction of current is shown with grey arrow.

If you notice the switches like diode, MOSFET, tristor, etc., have a port named "m". This is the "measurement" port and lets you see the switches current and voltage easily. For example, to see the MOSFET current/voltage you can use the diagram shown in Fig. 2.78. A demultiplexer block (Fig. 2.79) is used in Fig. 2.78 to guide the measured current and voltage to different scopes. As you see, the demultiplexer used in Fig. 2.78 has two outputs. The upper one is measured current and the lower one is measured voltage. MOSFET switch's current and voltage is shown in Figs. 2.80 and 2.81, respectively.

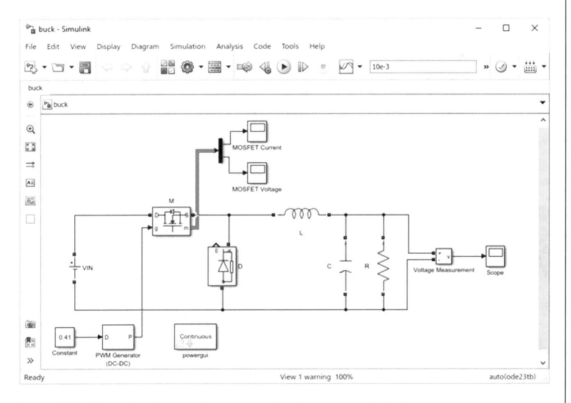

Figure 2.78: Measuring the switch current and voltage using the "m" port.

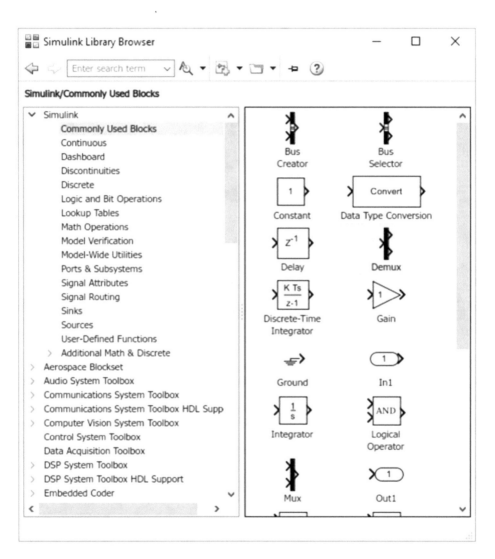

Figure 2.79: "Demux" block.

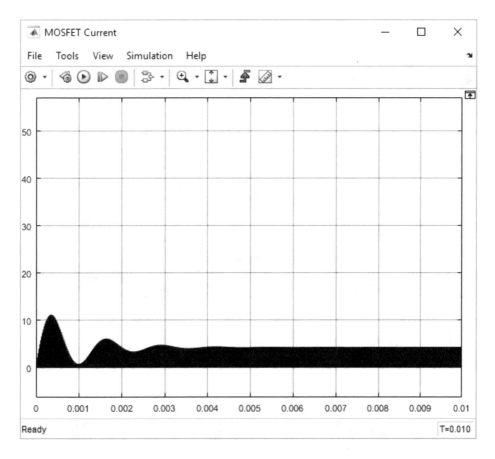

Figure 2.80: MOSFET "M" current.

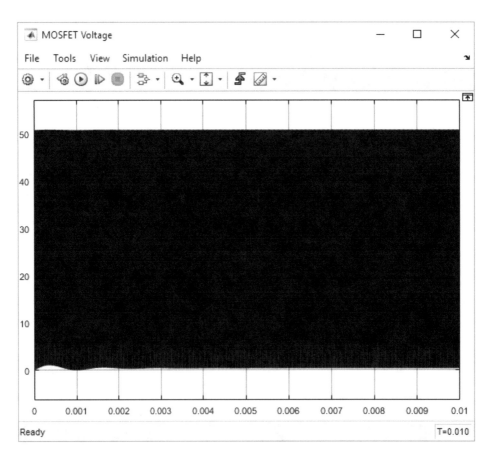

Figure 2.81: MOSFET "M" voltage.

2.4   EFFECT OF INPUT VOLTAGE CHANGES ON OUTPUT VOLTAGE OF A BUCK CONVERTER

We study the effect of input voltage changes on the output voltage of a Buck converter. To do this, we remove the input source by selecting the "VIN" block and press "Delete" on keyboard (Fig. 2.82). We add a "Controlled Voltage Source" block (Fig. 2.83) to the simulation file, as shown in Fig. 2.84. You can produce the required voltage to stimulate your circuit with the aid of "Controlled Voltage Source" block. For example, in order to produce a sudden change in voltage we use connect the "Controlled Voltage Source" block to a "Step" block. You can find the "Step" block in "Sources" section of Simulink®, as shown in Fig. 2.85. Step block settings are shown in Fig. 2.86. With this setting you obtain the waveform shown in Fig. 2.87.

Simulation result (load resistor voltage) is shown in Fig. 2.88. As can be seen, the output voltage changed (increased) at $t = 5$ ms since converter expects the 50 V input voltage but the input voltage is no longer 50 V for $t \geq 5$ ms. If constant output is required, a controller must be used. A controller set the duty cycle of MOSFET gate signal so the output keeps constant despite of disturbances.

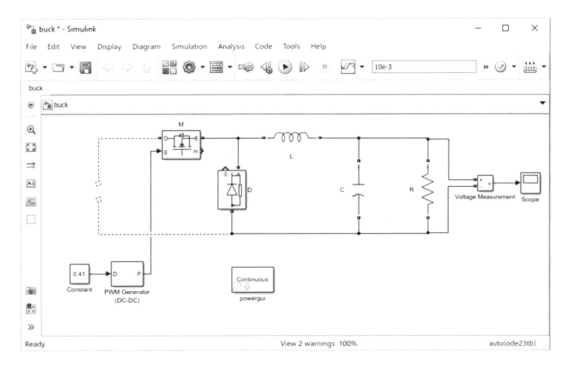

Figure 2.82: Removing the input voltage source.

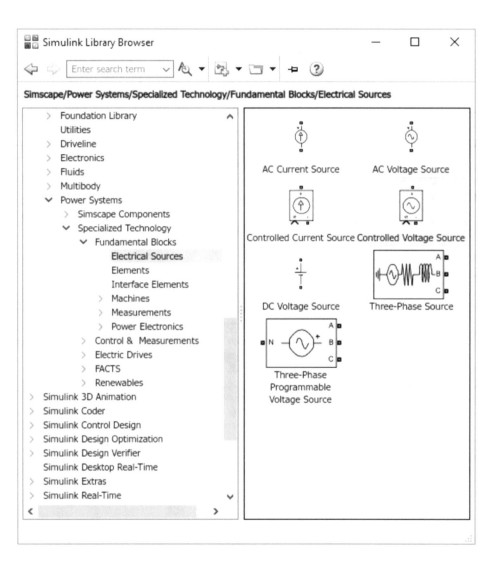

Figure 2.83: "Controlled Voltage Source" block.

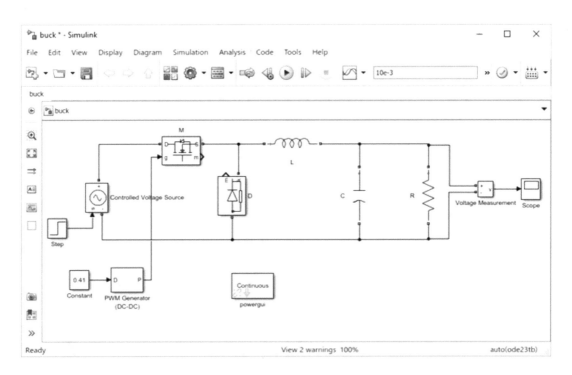

Figure 2.84: Simulation diagram to study the effect of step change in input voltage on the output voltage.

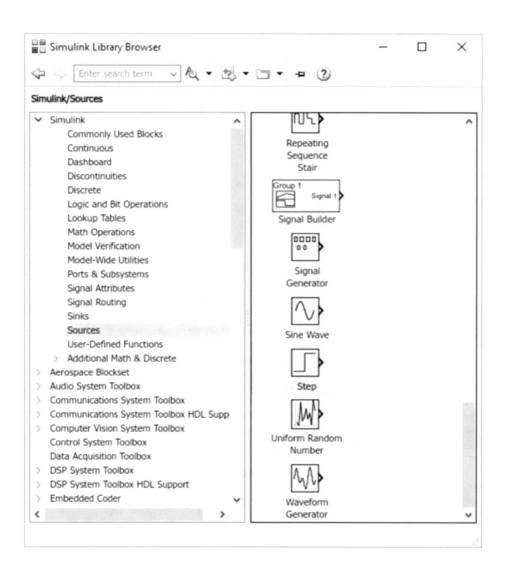

Figure 2.85: "Step" block.

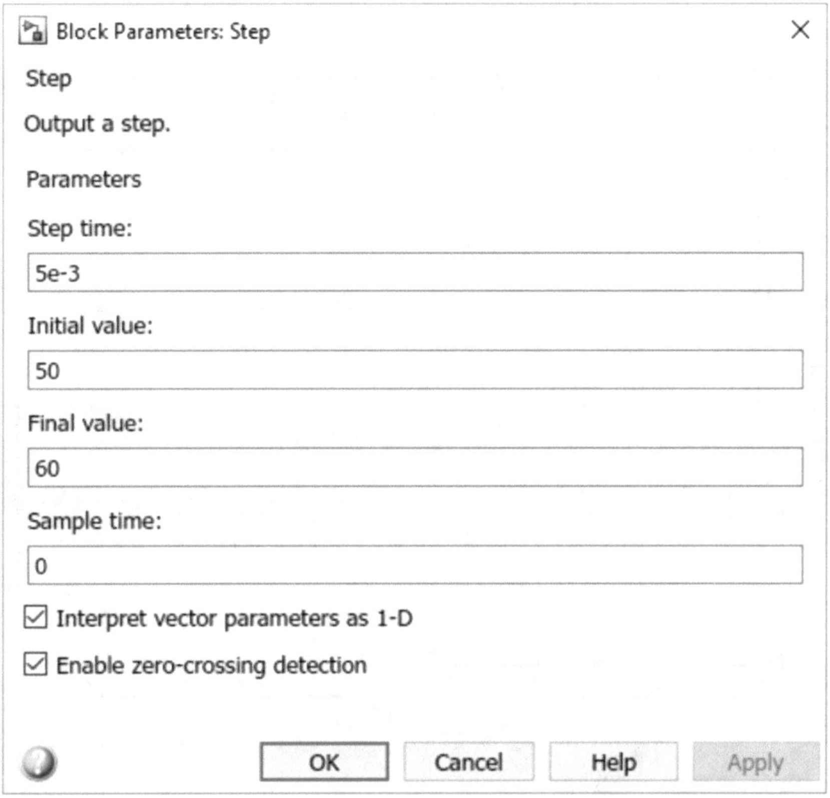

Figure 2.86: "Step" block setting.

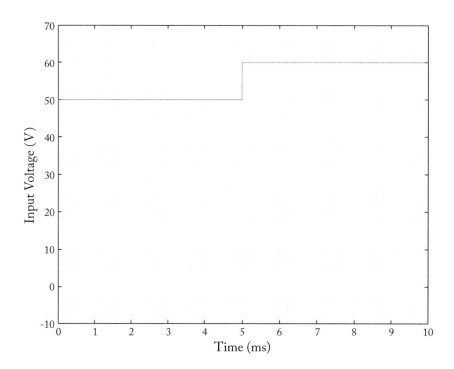

Figure 2.87: Input voltage waveform.

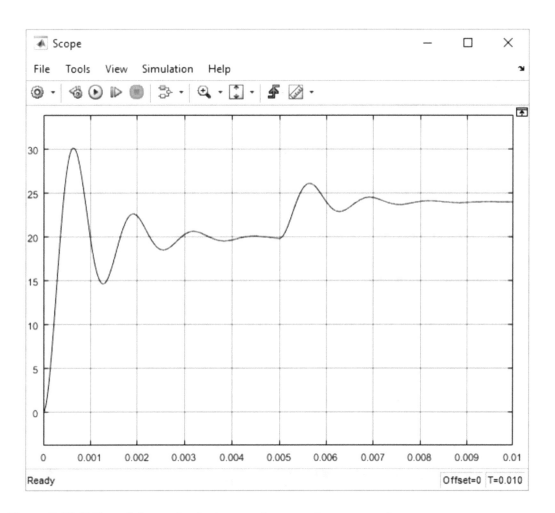

Figure 2.88: Effect of change in the input voltage on the output voltage.

2.5   EFFECT OF OUTPUT LOAD CHANGES ON OUTPUT VOLTAGE OF A BUCK CONVERTER

We want to study the effect of sudden change in output load on the output voltage of a Buck converter in this section. We modify the simulation file, as shown in Fig. 2.89. MOSFET "M1" is closed with the aid of "Step1" block. "Step1" block setting is shown in Fig. 2.90. According to the "Step1" settings, MOSFET "M1" is closed for $t \geq 7$ ms so the output load is $\frac{R_1 \times R}{R_1 + R}$. Resistors R_1 and R's value are 1 Ω and 5 Ω, respectively. So, for $t \geq 7$ ms output load is $\frac{1 \times 5}{1 + 5} = 0.83$ Ω. The simulation result is shown in Fig. 2.91. As can be seen, the load voltage is changed again.

All the inductors and capacitors include some Equivalent Series Resistance (ESR) which dissipate energy. If we place 0.1 Ω resistors in series with inductor and capacitor to model the ESR, the change in output voltage (Fig. 2.92) becomes larger in comparison with Fig. 2.91.

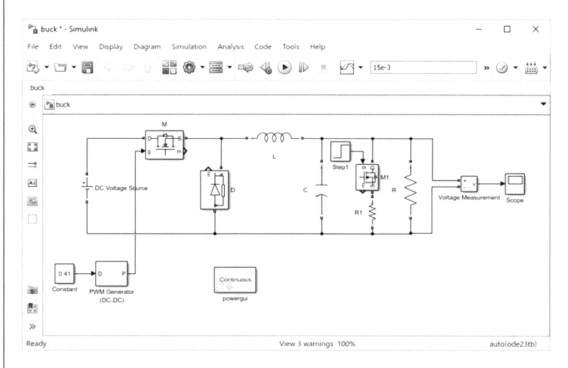

Figure 2.89: Simulation diagram to study the effect of output load changes on the output voltage.

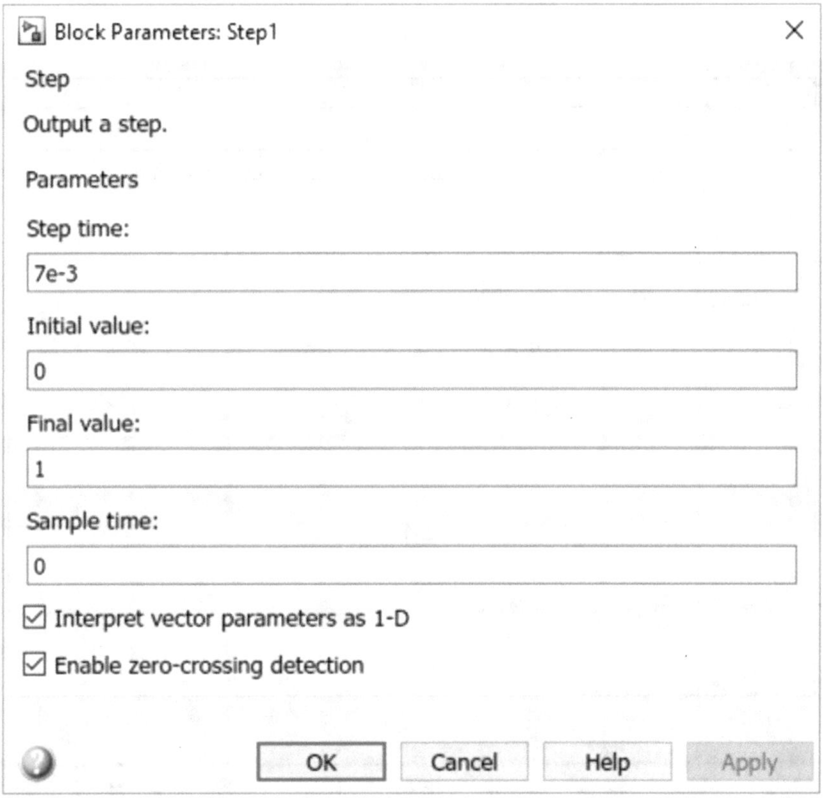

Figure 2.90: "Step1" block settings.

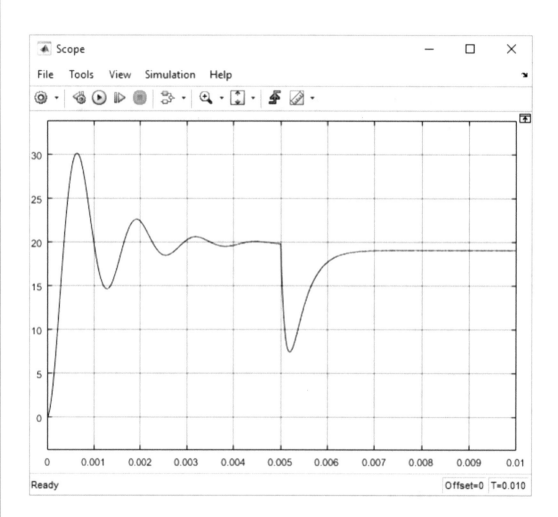

Figure 2.91: Effect of step change in output load on the output voltage.

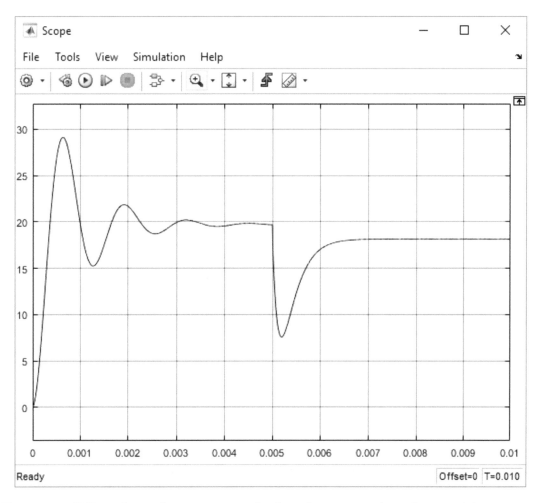

Figure 2.92: Effect of step change in output load on the output voltage for a model containing the elements ESR.

2.6   CONCLUSION

In this chapter we showed the importance of control in DC-DC converters with the aid of simulations. The output of a system without control changes in presence of disturbances.

We will study methods to keep the output constant despite of disturbances in future chapters.

REFERENCES

[1] *Simscape User's Guide*, MathWorks, 2015.

CHAPTER 3

Dynamics of DC-DC Converters

3.1 INTRODUCTION

In this chapter we try to extract a Linear Time Invariant (LTI) model for DC-DC converters. DC-DC converters are variable structure nonlinear dynamical systems. So, an LTI model is only an approximation. Although the obtained model is only an approximation, it is good enough to start the practical controller design process. One may use advanced control design techniques like H_∞ to consider the approximate nature of the extracted LTI model.

In this chapter we will introduce the State Space Averaging (SSA) which is one of the most important tools to model DC-DC converters working in Continuous Current Mode (CCM). We study the Buck converter as our first example. For ease of understanding, all details are shown. If you understand the Buck converter analysis well, you easily can apply the method to other types of converters. It is important that you make sure to spend sufficient time studying the first example carefully. We show how MATLAB®can do the mathematic machinery behind the SSA as well.

A graphical software tool to extract the model of common DC-DC converters is introduced here. Using this software, the modeling can be done automatically. The user need only enter the values to the software and the rest of the process is done by the software.

SSA is not a good tool to model converters working in Discontinuous Current Mode (DCM) however, a computer method to model converters working in DCM is introduced at the end of chapter. Refer to the references at the end of the chapter to learn the modeling procedure of converters working in DCM.

3.2 OVERVIEW OF STATE SPACE AVERAGING (SSA)

Assume we want to compare two students using their marks, as shown in Table 3.1.

In order to compare them fairly we must consider all the marks. This is done with the aid of averaging:

$$\frac{75 \times 4 + 70 \times 3 + 55 \times 2}{4 + 3 + 2} = 68.89$$

$$\frac{80 \times 4 + 65 \times 3 + 60 \times 2}{4 + 3 + 2} = 70.55.$$

Table 3.1: Students marks.

Student	Math (4 credits)	Physics (3 credits)	Biology (2 credits)
A	75	70	55
B	80	65	60

As can be seen, the second student is more successful since s/he has a higher average. Please note each mark is multiplied by the credits, so importance of the courses is entered into the averaging process.

The logic behind SSA is similar to the logic behind averaging marks. In this case, we average dynamic equations instead of marks. Studying an example is quite helpful. Details and mathematic machinery are shown in the next section.

Assume a simple Buck converter like that shown in Fig. 3.1. Based on the MOSFET status (on or off), two equivalent sub-circuits can be extracted, as shown in Fig. 3.2.

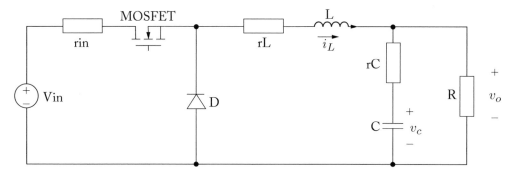

Figure 3.1: Buck converter.

We want to find a model for the Buck converter shown in Fig. 3.1 but we have two sub-circuits. Each sub-circuit has its own dynamic equation. We must find a way to average these two sets of equations.

Assume that the Buck converter of Fig. 3.1 spends 80% of the switching period in the MOSFET on state (Fig. 3.2a) and only 20% of switching period in the MOSFET off state (Fig. 3.2b). In this case, the Buck converter spends most of its time in the MOSFET on state. So, it is logical to give a higher weight to MOSFET on the equation set when averaging is done. It is not logical to average the equations (MOSFET on and MOSFET off) using the same weight.

SSA uses the percentage of switching time as the weights (credits in mark averaging problem). For instance, if the Buck converter spends 80% of switching time in the MOSFET on state

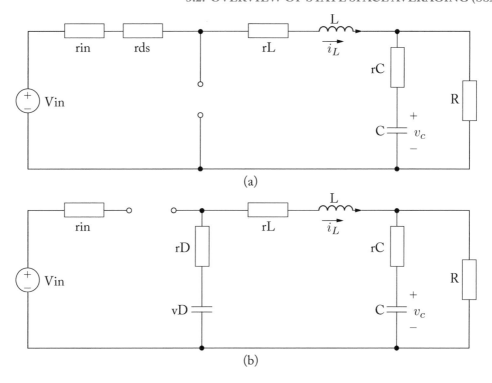

Figure 3.2: (a) Equivalent circuit for closed MOSFET and (b) equivalent circuit for opened MOSFET.

and 20% of switching time in the MOSFET off state, then the MOSFET on the equation set is multiplied by 0.8 and MOSFET off equation set is multiplied by 0.2.

We need an LTI model of the converter. Therefore, linearization must be applied to the obtained averaged equation set. As you remember from basic mathematic courses, $f(x_0 + \Delta x) \approx f(x_0) + f'(x_0)\Delta x$. We use the Taylor series to linearize the averaged equation set around the operating point.

Hence, we can summarize the steps of SSA as follows:

1. Dynamical equation of all the switch on and switch off states are extracted.

2. Equations are averaged using the duty ratio as weight.

3. Averaged equations are linearized around the operating point using the Taylor theorem.

In this chapter, we use the capital letters for steady state values and the tilde for small signal perturbations. Small signal perturbation is smaller than the steady state part. For example, the duty ratio of MOSFET gate signal is shown as $d = D + \tilde{d}$. This shows that the duty ratio of

MOSFET gate signal (d) is composed of two parts: steady state part (D) and small signal part ($\tilde{d}$) and $\tilde{d} \ll D$.

3.3   ILLUSTRATIVE EXAMPLE: BUCK CONVERTER

Assume a Buck converter like that shown in Fig. 3.3. rin, rL, and rC show the input source internal resistance, the inductor series resistance, and the capacitor series resistance, respectively. We assume values of the elements are selected such that a converter works in Continuous Current Mode (CCM).

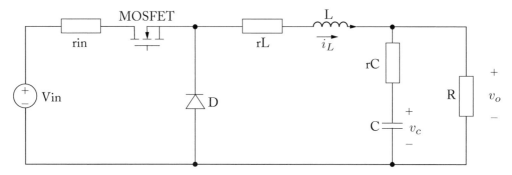

Figure 3.3: Buck converter circuit.

MOSFET is closed and opened with the aid of pulses shown in Fig. 3.4. When gate pulse is high (i.e., high logic level), MOSFET is closed. According to the Fig. 3.4, MOSFET is closed for $d \times T$ seconds and is opened for $T - d \times T = (1 - d) \times T$ seconds. T and d shows switching period and duty ratio, respectively.

Figure 3.4: MOSFET gate pulses.

Figures 3.5 and 3.6 show equivalent circuits for closed and opened MOSFET, respectively. When MOSFET is closed, diode D is reverse biased. When MOSFET is opened, diode D is forward biased. rds shows the MOSFET's on resistance (i.e., the resistance between drain and source).

These two linear circuits are used to obtain the converter's transfer function. First, each circuit is analyzed separately. After that, obtained equations are superimposed to obtain a unique transfer function.

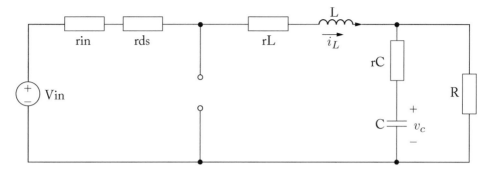

Figure 3.5: Equivalent circuit for closed MOSFET.

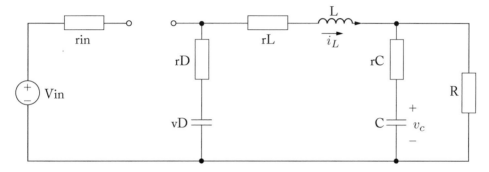

Figure 3.6: Equivalent circuit for opened MOSFET. *rD* and *vD* shows the diode resistance and forward voltage drop, respectively.

3.4   CLOSED MOSFET CASE

According to Fig. 3.5, MOSFET is closed when gate pulses are high. Since gate pulses are periodic this happens when $n.T < t < n.T + d.T$, $n = 0, 1, 2, 3, \ldots$. So, MOSFET is closed for $d.T$ seconds.

When MOSFET is closed, circuit equation can be written as:

$$\begin{cases} (r_{in} + r_{ds} + r_L)\, i_L + L\dfrac{di_L}{dt} + R\left(i_L - C\dfrac{dv_C}{dt}\right) = v_{in} \\ r_C C\dfrac{dv_C}{dt} + v_C = R\left(i_L - C\dfrac{dv_C}{dt}\right). \end{cases} \quad (3.1)$$

Equations can be simplified using simple algebraic manipulations,

$$\begin{cases} (r_{in} + r_{ds} + r_L + R)\, i_L + \dfrac{di_L}{dt} - RC\dfrac{dv_C}{dt} = v_{in} \\ (r_C + R)\, C\dfrac{dv_C}{dt} + v_C = Ri_L \end{cases} \quad (3.2)$$

$$\begin{cases} (r_{in} + r_{ds} + r_L + R)\,i_L + L\dfrac{di_L}{dt} - \dfrac{R}{R + r_C}\,(Ri_L - v_C) = v_{in} \\ C\dfrac{dv_C}{dt} = \dfrac{1}{R + r_C}\,(Ri_L - v_C) \end{cases} \tag{3.3}$$

$$\begin{cases} L\dfrac{di_L}{dt} = -(r_{in} + r_{ds} + r_L + R)\,i_L + \dfrac{R^2}{R+r_C}\,i_L - \dfrac{R}{R + r_C}\,v_C + v_{in} \\ C\dfrac{dv_C}{dt} = \dfrac{1}{R + r_C}\,(Ri_L - v_C)\,. \end{cases} \tag{3.4}$$

Output equation[1] can be written as

$$v_o = R\left(i_L - C\frac{dv_C}{dt}\right) = R\left(\frac{r_C}{r_C + R}i_L + \frac{1}{R + r_C}v_C\right). \tag{3.5}$$

3.5 OPEN MOSFET CASE

According to Fig. 3.6, MOSFET is opened when gate pulses are low (i.e., logic level zero). Since gate pulses are periodic this happens when $n.T + d.T < t < (n + 1).T + d.T$, $n = 0, 1, 2, 3, \ldots$. So, MOSFET is opened for $(1 - d).T$ seconds.

When MOSFET is opened, the circuit equation can be written as:

$$\begin{cases} V_D + (r_D + r_L)\,i_L + L\dfrac{di_L\,(t)}{dt} + r_C C\dfrac{dv_C}{dt} + v_C = 0 \\ r_C C\dfrac{dv_C}{dt} + v_C = R\left(i_L - C\dfrac{dv_C}{dt}\right). \end{cases} \tag{3.6}$$

Equations can be simplified using simple algebraic manipulations:

$$\begin{cases} V_D + (r_D + r_L)\,i_L + L\dfrac{di_L\,(t)}{dt} + \dfrac{R.r_C}{R + r_C}\,i_L + \dfrac{R}{R + r_C}\,v_C = 0 \\ C\dfrac{dv_C}{dt} = \dfrac{R}{R + r_C}\,i_L - \dfrac{1}{R + r_C}\,v_C\,. \end{cases} \tag{3.7}$$

$$\begin{cases} L\dfrac{di_L\,(t)}{dt} = -\left(r_D + r_L + \dfrac{R.r_C}{R + r_C}\right)i_L - \dfrac{R}{R + r_C}\,v_C - V_D \\ C\dfrac{dv_C}{dt} = \dfrac{R}{R + r_C}\,i_L - \dfrac{1}{R + r_C}\,v_C\,. \end{cases} \tag{3.8}$$

Output equation can be written as

$$v_o = R\left(i_L - C\frac{dv_C}{dt}\right) = R\left(\frac{r_C}{r_C + R}i_L + \frac{1}{R + r_C}v_C\right). \tag{3.9}$$

[1]Generally, load voltage is taken as output.

3.6 AVERAGING

Results of previous analysis are superimposed using averaging. MOSFET is closed for $d.T$ seconds and is opened for $(1-d).T$ seconds. So, it is fair to multiply the obtained equation set by the length of period which is valid and averaged over one period.

Inductor current equations are obtained as:

$$\begin{cases} L\dfrac{di_L}{dt} = -\left(r_{in} + r_{ds} + r_L + R - \dfrac{R^2}{R+r_C}\right)i_L - \dfrac{R}{R+r_C}v_C + v_{in}, \ n.T < t < n.T + d.T \\[4mm] L\dfrac{di_L(t)}{dt} = -\left(r_D + r_L + \dfrac{R.r_C}{R+r_C}\right)i_L - \dfrac{R}{R+r_C}v_C - V_D, \quad n.T + d.T < t < (n+1)T. \end{cases}$$

$$(3.10)$$

Left and right sides of the equations are multiplied by the length of the time interval

$$\begin{cases} L\dfrac{di_L}{dt} \times d.T = \left(-(r_{in} + r_{ds} + r_L + R)i_L + \dfrac{R^2}{R+r_C}i_L - \dfrac{R}{R+r_C}v_C + v_{in}\right) \times d.T \\[4mm] L\dfrac{di_L(t)}{dt} \times (1-d).T = \left(-\left(r_D + r_L + \dfrac{R.r_C}{R+r_C}\right)i_L - \dfrac{R}{R+r_C}v_C - V_D\right) \times (1-d).T. \end{cases}$$

$$(3.11)$$

Corresponding sides are added together:

$$d.T \times L\dfrac{di_L}{dt} + (1-d).T \times L\dfrac{di_L}{dt}$$
$$= d.T \times \left(-(r_{in} + r_{ds} + r_L + R)i_L + \dfrac{R^2}{R+r_C}i_L - \dfrac{R}{R+r_C}v_C + v_{in}\right) \qquad (3.12)$$
$$+ (1-d).T \times \left(-\left(r_D + r_L + \dfrac{R.r_C}{R+r_C}\right)i_L - \dfrac{R}{R+r_C}v_C - V_D\right).$$

Averaging is realized by multiplying both sides by $\frac{1}{T}$:

$$d \times L\dfrac{di_L}{dt} + (1-d) \times L\dfrac{di_L}{dt}$$
$$= d \times \left(-(r_{in} + r_{ds} + r_L + R)i_L + \dfrac{R^2}{R+r_C}i_L - \dfrac{R}{R+r_C}v_C + v_{in}\right)$$
$$+ (1-d) \times \left(-\left(r_D + r_L + \dfrac{R.r_C}{R+r_C}\right)i_L - \dfrac{R}{R+r_C}v_C - V_D\right). \qquad (3.13)$$

After some simple algebraic manipulations,

$$L\dfrac{di_L}{dt} = -d \times \left(r_{in} + r_{ds} + r_L + R - \dfrac{R^2}{R+r_C}\right)i_L$$
$$- (1-d) \times \left(r_D + r_L + \dfrac{R.r_C}{R+r_C}\right)i_L - \dfrac{R}{R+r_C}v_C - (1-d)V_D + dv_{in} \qquad (3.14)$$

$$L\dfrac{di_L}{dt} = -d \times R_1 i_L - (1-d) \times R_2 i_L - \dfrac{R}{R+r_C}v_C - (1-d)V_D + dv_{in}$$

$$R_1 = r_{in} + r_{ds} + r_L + R - \frac{R^2}{R + r_C} \qquad (3.15)$$

$$R_2 = r_D + r_L + \frac{R.r_C}{R + r_C} \qquad (3.16)$$

is obtained.

The same procedure can be applied to the capacitor voltage equations:

$$\begin{cases} C\dfrac{dv_C}{dt} = \dfrac{R}{R + r_C}i_L - \dfrac{1}{R + r_C}v_C & n.T < t < n.T + d.T \\[2mm] C\dfrac{dv_C}{dt} = \dfrac{R}{R + r_C}i_L - \dfrac{1}{R + r_C}v_C & n.T + d.T < t < (n + 1)T \end{cases} \qquad (3.17)$$

$$d \times C\frac{dv_C}{dt} + (1 - d) \times C\frac{dv_C}{dt}$$
$$= d \times \left(\frac{R}{R + r_C}i_L - \frac{1}{R + r_C}v_C \right) + (1 - d) \left(\frac{R}{R + r_C}i_L - \frac{1}{R + r_C}v_C \right)$$
$$C\frac{dv_C}{dt} = \left(\frac{R}{R + r_C}i_L - \frac{1}{R + r_C}v_C \right). \qquad (3.18)$$

Hence, the equation of *average* system can be written as:

$$\begin{cases} L\dfrac{di_L}{dt} = -d \times R_1 i_L - (1 - d) \times R_2 i_L - \dfrac{R}{R + r_C}v_C - (1 - d)V_D + dv_{in} \\[2mm] C\dfrac{dv_C}{dt} = \dfrac{R}{R + r_C}i_L - \dfrac{1}{R + r_C}v_C \end{cases}, \qquad (3.19)$$

where

$$R_1 = r_{in} + r_{ds} + r_L + R - \frac{R^2}{R + r_C}$$

$$R_2 = r_D + r_L + \frac{R.r_C}{R + r_C}.$$

This average system can be used to obtain the steady state currents and voltages. To obtain the steady state currents and voltages one must replace the left-hand side with zero. Capital letters show the steady state values. For instance, I_L shows the steady-state inductor current:

$$\begin{cases} 0 = -D \times R_1 I_L - (1 - D) \times R_2 I_L + \dfrac{R}{R + r_C}V_C - (1 - D)V_D + DV_{IN} \\[2mm] 0 = \dfrac{R}{R + r_C}I_L - \dfrac{1}{R + r_C}V_C. \end{cases} \qquad (3.20)$$

Steady state values are found as:

$$\begin{cases} I_L = \dfrac{((R + r_C)(DV_{IN} - (1-D)V_D)}{(R + r_C)R_2 + R^2 + D(R + r_C)(R_1 - R_2)} \\ V_C = \dfrac{((R + r_C)(DV_{IN} - (1-D)V_D)}{(R + r_C)R_2 + (1-2D)R^2 + D(R + r_C)(R_1 - R_2)} \times R. \end{cases} \tag{3.21}$$

One can use MATLAB®to solve equation set (3.20). To do this, the following code can be used:

```
clc
clear all

syms R1 R2 R D IL VC rC rL VD vIN

eq1=-D*R1*IL-(1-D)*R2*IL-R/(R+rC)*VC-(1-D)*VD+D*vIN;
eq2=R/(R+rC)*IL-1/(R+rC)*VC;

DC_operatingPoint=solve(eq1,eq2,'[IL VC]');

disp('IL=')
pretty(simplify(DC_operatingPoint.IL))

disp('VC=')
pretty(simplify(DC_operatingPoint.VC))
```

If we ignore rin, rds, rD, VD (i.e., $rin = rds = rD = VD = 0$) steady state values are obtained as:

$$\begin{cases} I_L = \dfrac{D \times V_{IN}}{R} \\ V_C = D \times V_{IN}. \end{cases} \tag{3.22}$$

Which is the familiar equations of an ideal (i.e., a converter with 100% efficiency) Buck converter operating in CCM.

The averaging procedure must be applied to the output equation as well:

$$dT \times v_o + (1-d)T \times v_o$$
$$= dT \times R\left(\frac{r_C}{r_C + R}i_L + \frac{1}{R + r_C}v_C\right) + (1-d)T \tag{3.23}$$
$$\times R\left(\frac{r_C}{r_C + R}i_L + \frac{1}{R + r_C}v_C\right).$$

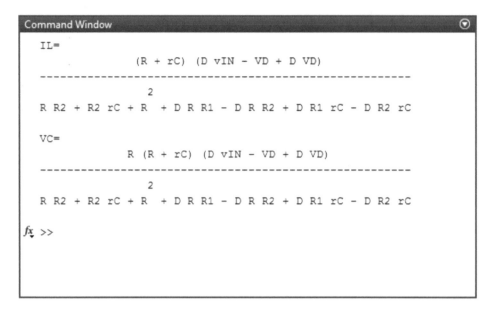

Figure 3.7: The steady state current and voltage.

Averaging the equation set is complete with multiplying both sides by $\frac{1}{T}$. If we multiply both sides by $\frac{1}{T}$ and do the simple algebraic manipulations,

$$v_o = R \left(\frac{r_C}{r_C + R} i_L + \frac{1}{R + r_C} v_C \right) \tag{3.24}$$

is obtained.

3.7   LINEARIZATION OF THE EQUATIONS

Systems average equations are obtained as:

$$\begin{cases} L \dfrac{di_L}{dt} = -d \times R_1 i_L - (1-d) \times R_2 i_L - \dfrac{R}{R + r_C} v_C - (1-d) V_D + d v_{in} \\ C \dfrac{dv_C}{dt} = \dfrac{R}{R + r_C} i_L - \dfrac{1}{R + r_C} v_C \end{cases}, \tag{3.25}$$

where

$$R_1 = r_{in} + r_{ds} + r_L + R - \frac{R^2}{R + r_C}$$

$$R_2 = r_D + r_L + \frac{R \times r_C}{R + r_C}.$$

Linearization is done using the Taylor series. Assume that

$$i_L = I_L + \tilde{i}_L \tag{3.26}$$

$$v_C = V_C + \tilde{v}_C \tag{3.27}$$

$$d = D + \tilde{d}, \tag{3.28}$$

where $\tilde{i}_L \ll I_L$, $\tilde{v}_C \ll V_C$, and $\tilde{d} \ll D$. Diode-forward voltage drop has been assumed to be constant. These equations shows that each state has a steady state value (I_L, V_C, and D) and a small perturbation ($\tilde{i}_L$, $\tilde{v}_C$, and $\tilde{d}$). Steady state values are obtained by solving the average system equation with the left-hand side equal to zero (see Eq. (3.20)).

We place these new variables into the equations

$$L\frac{d\left(I_L + \tilde{i}_L\right)}{dt} = -(D + \tilde{d}) \times R_1(I_L + \tilde{i}_L) - \left(1 - \left(D + \tilde{d}\right)\right) \times R_2(I_L + \tilde{i}_L)$$
$$- \frac{R}{R + r_C}(V_C + \tilde{v}_C) - \left(1 - \left(D + \tilde{d}\right)\right) V_D + (D + \tilde{d})(V_{IN} + \tilde{v}_{in}) \tag{3.29}$$

$$C\frac{d(V_C + \tilde{v}_C)}{dt} = \frac{R}{R + r_C}(I_L + \tilde{i}_L) - \frac{1}{R + r_C}(V_C + \tilde{v}_C). \tag{3.30}$$

After simple algebraic manipulations in Eq. (3.29),

$$L\frac{d\left(I_L + \tilde{i}_L\right)}{dt} = -(D + \tilde{d}) \times R_1(I_L + \tilde{i}_L) - \left(1 - \left(D + \tilde{d}\right)\right) \times R_2(I_L + \tilde{i}_L)$$
$$- \frac{R}{R + r_C}(V_C + \tilde{v}_C) - \left(1 - \left(D + \tilde{d}\right)\right) V_D + (D + \tilde{d})(V_{IN} + \tilde{v}_{in}) \implies$$

$$L\frac{d\left(I_L + \tilde{i}_L\right)}{dt} = -R_1 D I_L - R_1 D \tilde{i}_L - R_1 I_L \tilde{d} - R_1 \tilde{i}_L \tilde{d} + R_2(D-1) I_L + R_2(D-1)\tilde{i}_L$$
$$+ R_2 I_L \tilde{d} + R_2 \tilde{i}_L \tilde{d} - \frac{R}{R + r_C} V_C - \frac{R}{R + r_C} \tilde{v}_c + (D-1) V_D + V_D \tilde{d} + D V_{IN}$$
$$+ D\tilde{v}_{in} + V_{IN}\tilde{d} + \tilde{v}_{in}\tilde{d} \implies$$

$$L\frac{d\left(I_L + \tilde{i}_L\right)}{dt} = -R_1 D I_L + R_2(D-1) I_L + (D-1) V_D - \frac{R}{R + r_C} V_C + D V_{IN} + \tilde{v}_{in}\tilde{d}$$
$$+ R_2 \tilde{i}_L \tilde{d} - R_1 \tilde{i}_L \tilde{d} + (R_2(D-1) - R_1 D)\tilde{i}_L - \frac{R}{R + r_C}\tilde{v}_c$$
$$+ (V_{IN} + V_D + (R_2 - R_1)I_L)\tilde{d} + D\tilde{v}_{in} \tag{3.31}$$

is obtained. $L\frac{d(I_L + \tilde{i}_L)}{dt} = L\frac{d(\tilde{i}_L)}{dt}$ since derivative of a constant term is zero. The right-hand side can be grouped into three groups:

- $-R_1 D I_L + R_2 (D - 1) I_L + (D - 1) V_D - \frac{R}{R+r_C} V_C + D V_{IN}$,

- $\tilde{v}_{in} \tilde{d} + R_2 \tilde{i}_L \tilde{d} - R_1 \tilde{i}_L \tilde{d}$, and

- $+(R_2 (D - 1) - R_1 D)\tilde{i}_L - \frac{R}{R+r_C} \tilde{v}_c + (V_{IN} + V_D + (R_2 - R_1)I_L)\tilde{d} + D \tilde{v}_{in}$.

If we place the steady state values obtained before (see Eq. (3.21)) into the $-R_1 D I_L + R_2 (D - 1) I_L + (D - 1) V_D - \frac{R}{R+r_C} V_C + D V_{IN}$, the result will become 0. Second group can be vanished as well. Because multiplication of two small numbers results in a small number too. So, only the third group is important:

$$L\frac{d(\tilde{i}_L)}{dt} \approx + (R_2 (D - 1) - R_1 D)\tilde{i}_L - \frac{R}{R + r_C}\tilde{v}_c$$
$$+ (V_{IN} + V_D + (R_2 - R_1)I_L)\tilde{d} + D \tilde{v}_{in}. \tag{3.32}$$

The same procedure can be applied to the capacitor voltage equation (Eq. (3.30)):

$$C\frac{d(V_C + \tilde{v}_C)}{dt} = \frac{R}{R + r_C}(I_L + \tilde{i}_L) - \frac{1}{R + r_C}(V_C + \tilde{v}_C) \implies$$
$$C\frac{d(V_C + \tilde{v}_C)}{dt} = \frac{R}{R + r_C}I_L - \frac{1}{R + r_C}V_C + \frac{R}{R + r_C}\tilde{i}_L - \frac{1}{R + r_C}\tilde{v}_C. \tag{3.33}$$

V_C is constant so its derivative is zero. So,

$$C\frac{d(\tilde{v}_C)}{dt} = \frac{R}{R + r_C}I_L - \frac{1}{R + r_C}V_C + \frac{\bullet R}{R + r_C}\tilde{i}_L - \frac{1}{R + r_C}\tilde{v}_C. \tag{3.34}$$

The right-hand side can be grouped into two groups:

- $\frac{R}{R+r_C} I_L - \frac{1}{R+r_C} V_C$

- $\frac{R}{R+r_C} \tilde{i}_L - \frac{1}{R+r_C} \tilde{v}_C$.

If we place the steady state values obtained before (see Eq. (3.21)) into the $\frac{R}{R+r_C} I_L - \frac{1}{R+r_C} V_C$ the result will be 0. The only important terms belong to the second group:

$$C\frac{d(\tilde{v}_C)}{dt} \approx \frac{R}{R + r_C}\tilde{i}_L - \frac{1}{R + r_C}\tilde{v}_C. \tag{3.35}$$

The output equation is linearized in the same way. Output voltage (v_o) is decomposed into two components:

- large signal component (V_o) and

- small signal component $(\tilde{v}_o)$.

We replace variables in the averaged equation (Eq. (3.24)) with their corresponding large signal and small signal parts:

$$v_o = R\left(\frac{r_C}{r_C + R}i_L + \frac{1}{R + r_C}v_C\right) \Longrightarrow$$

$$V_o + \tilde{v}_o = R\left(\frac{r_C}{r_C + R}(I_L + \tilde{i}_L) + \frac{1}{R + r_C}(V_C + \tilde{v}_C)\right) \Longrightarrow$$

$$V_o + \tilde{v}_o = \frac{R.r_C}{r_C + R}I_L + \frac{R}{R + r_C}V_C + \frac{R.r_C}{r_C + R}\tilde{i}_L + \frac{R}{R + r_C}\tilde{v}_C. \tag{3.36}$$

The output's large signal part can be obtained as:

$$V_o = \frac{R.r_C}{r_C + R}I_L + \frac{R}{R + r_C}V_C. \tag{3.37}$$

If we replace the steady state values for I_L and V_C (Eq. (3.21)) into (3.37) we obtain the output steady state value.

We are interested in the small signal part. So, only the small signal terms are considered:

$$\tilde{v}_o = \frac{R.r_C}{r_C + R}\tilde{i}_L + \frac{R}{R + r_C}\tilde{v}_C. \tag{3.38}$$

Therefore, the linearized small signal model of a Buck converter can be written as:

$$\begin{cases} \dfrac{d(\tilde{i}_L)}{dt} \approx \dfrac{1}{L}\left[(R_2(D-1) - R_1 D)\tilde{i}_L - \dfrac{R}{R + r_C}\tilde{v}_C + (V_{IN} + V_D + (R_2 - R_1)I_L)\tilde{d} + D\tilde{v}_{in}\right] \\ \dfrac{d(\tilde{v}_C)}{dt} \approx \dfrac{1}{C}\left[\dfrac{R}{R + r_C}\tilde{i}_L - \dfrac{1}{R + r_C}\tilde{v}_C\right] \end{cases}$$

$$\tilde{v}_o = \frac{R.r_C}{r_C + R}\tilde{i}_L + \frac{R}{R + r_C}\tilde{v}_C$$

$$R_1 = r_{in} + r_{ds} + r_L + R - \frac{R^2}{R + r_C}$$

$$R_2 = r_D + r_L + \frac{R \times r_C}{R + r_C}. \tag{3.39}$$

It can be written in the form of a state space equation:

$$\begin{cases} \dot{x} = Ax + Bu \\ y = \mathbb{C}x, \end{cases} \tag{3.40}$$

where

$$x = \begin{bmatrix} \tilde{i}_L \\ \tilde{v}_c \end{bmatrix},$$

$$u = \begin{bmatrix} \tilde{d} \\ \tilde{v}_{in} \end{bmatrix},$$

$$y = v_o,$$

$$A = \begin{bmatrix} \dfrac{R_2(D-1) - R_1 D}{L} & -\dfrac{R}{(R+r_C)L} \\ \dfrac{R}{(R+r_C)C} & -\dfrac{1}{(R+r_C)C} \end{bmatrix},$$

$$B = \begin{bmatrix} \dfrac{(V_{IN} + V_D + (R_2 - R_1)I_L)}{L} & \dfrac{D}{L} \\ 0 & 0 \\ \dfrac{}{C} & \dfrac{}{C} \end{bmatrix},$$

$$\mathbb{C} = \begin{bmatrix} \dfrac{R.r_C}{r_C + R} & \dfrac{R}{r_C + R} \end{bmatrix}, \tag{3.41}$$

$\mathbb{C}$ is a matrix and it must not be confused with capacitor value C.

MATLAB®is quite helpful to do the calculations. Assume a Buck converter with the following parameters: $R = 5\,\Omega$, $V_{in} = 50$ V, $r_{in} = 0.1\,\Omega$, $L = 400\,\mu$H, $r_L = 0.1\,\Omega$, $C = 100\,\mu$F, $r_C = 0.05\,\Omega$, $D = 0.41$, $r_{ds} = 0.1\,\Omega$, $r_D = 0.1\,\Omega$, and $V_D = 0.7$ V.

The following program calculates the small signal transfer functions for the given values. After running the program, $\dfrac{\tilde{v}_o(s)}{\tilde{d}(s)} = \dfrac{6184s + 1.237 \times 10^9}{s^2 + 2574s + 2.568 \times 10^7}$ and $\dfrac{\tilde{v}_o(s)}{\tilde{v}_{in}(s)} = \dfrac{50.74s + 1.015 \times 10^7}{s^2 + 2574s + 2.568 \times 10^7}$ are obtained. The Bode plot of these transfer functions is shown in Figs. 3.8 and 3.9.

```
%This program calculate the small signal transfer
%functions for Buck converter
R=5;

VIN=50;
rin=.1;

L=400e-6;
rL=.1;

C=100e-6;
rC=.05;
```

```
rD=.01;
VD=.7;

rds=.1;

D=.41;

R1=rin+rds+rL+R*rC/(R+rC);
R2=rD+rL+R*rC/(R+rC);

IL=(R+rC)*(D*VIN-(1-D)*VD)/((R+rC)*R2+R^2+D*(R+rC)*(R1-R2));

A=[(R2*(D-1)-R1*D)/L -R/(R+rC)/L;R/(R+rC)/C -1/(R+rC)/C];
B=[(VIN+VD+(R2-R1)*IL)/L D/L;0 0];
%C shows the capacitance so CC is used for matrix
CC=[R*rC/(rC+R) R/(R+rC)];
H=tf(ss(A,B,CC,0));
% transfer function between output voltage and duty ratio
vO_d=H(1)
%transfer function between output voltage and input source
vO_vin=H(2)
figure(1)
bode(vO_d), grid on
figure(2)
bode(vO_vin), grid on
```

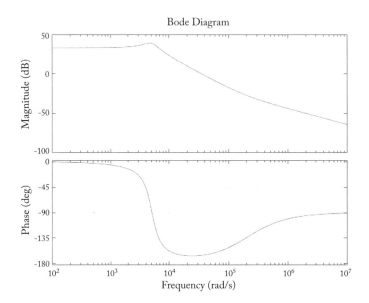

Figure 3.8: Bode plot of vO_d $= \frac{\tilde{v}_o(s)}{\tilde{d}(s)}$ for a Buck converter.

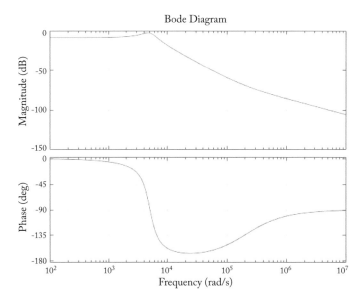

Figure 3.9: Bode plot of vO_vin $= \frac{\tilde{v}_o(s)}{\tilde{v}_{in}(s)}$ for a Buck converter.

3.8   OBTAINING THE SMALL SIGNAL TRANSFER FUNCTIONS USING MATLAB®

MATLAB®can do the machinery easily without any error. The following program can extract the small signal transfer function for a Buck converter. Outputs are shown in Fig. 3.10. As you see, results are the same with the previous analysis.

```
% This program extract the small signal transfer function
clc
clear all;
% Elements values
R=5;       %Load resistor
VIN=50;    %Input source voltage
rin=.1;    %Input source internal resistance
L=400e-6;  %inductor
rL=.1;     %inductor series resistance
C=100e-6;  %capacitor
rC=.05;    %capacitor series resistance
rD=.01;    %Diode series resistance
VD=.7;     %Diode forward voltage drop
rds=.1;    %MOSFET on resistance
D=.41;     %Duty ratio

% Symbolic variables
%iL:  inductor current
%vC:  capacitor voltage
%vin: input voltage source
%vD:  diode forward voltage drop
%d:   duty cycle
syms iL vC vin vD d

%CLOSED MOSFET EQUATIONS
%d(iL)/dt for closed MOSFET
M1=(-(rin+rds+rL+(R*rC/(R+rC)))*iL-R/(R+rC)*vC+vin)/L;
%d(vC)/dt for closed MOSFET
M2=(R/(R+rC)*iL-1/(R+rC)*vC)/C;
vO1=R*(rC/(rC+R)*iL+1/(R+rC)*vC);
%OPENED MOSFET EQUATIONS
%d(iL)/dt for opened MOSFET
M3=(-(rD+rL+R*rC/(R+rC))*iL-R/(R+rC)*vC-vD)/L;
```

```
%%d(vC)/dt for opened MOSFET
M4=(R/(R+rC)*iL-1/(R+rC)*vC)/C;
vO2=R*(rC/(rC+R)*iL+1/(R+rC)*vC);
%AVERAGING
MA1= simplify(d*M1+(1-d)*M3);
MA2= simplify(d*M2+(1-d)*M4);
vO= simplify(d*vO1+(1-d)*vO2);
% DC OPERATING POINT CALCULATION
MA_DC_1=subs(MA1,[vin vD d],[VIN VD D]);
MA_DC_2=subs(MA2,[vin vD d],[VIN VD D]);

DC_SOL= solve(MA_DC_1==0,MA_DC_2==0,'iL','vC');

%IL is the inductor current steady state value
IL=eval(DC_SOL.iL);
%VC is the capacitor current steady state value
VC=eval(DC_SOL.vC);

%LINEARIZATION
% .
% x=Ax+Bu
%vector x=[iL;vC] is assumed. vector x is states.
%u=[vin;d] where vin=input voltage source and d=duty.
%vector u is system inputs.
%
A11=subs(simplify(diff(MA1,iL)),[iL vC d vD],[IL VC D VD]);
A12=subs(simplify(diff(MA1,vC)),[iL vC d vD],[IL VC D VD]);

A21=subs(simplify(diff(MA2,iL)),[iL vC d vD],[IL VC D VD]);
A22=subs(simplify(diff(MA2,vC)),[iL vC d vD],[IL VC D VD]);

%variable A is matrix A in state space equation
A=eval([A11 A12; A21 A22]);

B11=subs(simplify(diff(MA1,vin)),[iL vC d vD vin],
         [IL VC D VD VIN]);
B12=subs(simplify(diff(MA1,d)),[iL vC d vD vin],
         [IL VC D VD VIN]);
```

```
B21=subs(simplify(diff(MA2,vin)),[iL vC d vD vin],
         [IL VC D VD VIN]);
B22=subs(simplify(diff(MA2,d)),[iL vC d vD vin],
         [IL VC D VD VIN]);


% variable B is matrix B in state space equation
B=eval([B11 B12; B21 B22]);

CC1=subs(simplify(diff(vO,iL)),[iL vC d vD],[IL VC D VD]);
CC2=subs(simplify(diff(vO,vC)),[iL vC d vD],[IL VC D VD]);
%variable CC is matrix C in state space equation
CC=eval([CC1 CC2]);
% variable D shows duty so DD is used.
DD11=subs(simplify(diff(vO,vin)),[iL vC d vD vin],
          [IL VC D VD VIN]);
DD12=subs(simplify(diff(vO,d)),[iL vC d vD vin],
          [IL VC D VD VIN]);

% variable DD is matrix D in state space equation
% variable D shows duty so DD is used.
DD=eval([DD11 DD12]);

H=tf(ss(A,B,CC,DD));

%transfer function between input source and load resistor voltage
% ~
vR_vin=H(1,1)  % vR(s)
% ----
% ~
% vin(s)

%transfer function between duty ratio and load resistor voltage
% ~
vR_d=H(1,2)     %vR(s)
%----
% ~
%d(s)
```

```
Command Window                                                          ⊙
                                                                        ^
         50.74 s + 1.015e07
     --------------------------
     s^2 + 2574 s + 2.568e07


     Continuous-time transfer function.



     vR_d =

          6184 s + 1.237e09
     --------------------------
     s^2 + 2574 s + 2.568e07


     Continuous-time transfer function.

fx >>                                                                   ∨
```

Figure 3.10: Calculated transfer functions.

The program is composed of three parts:

- taking the parameter values;

- taking the converters dynamical equation, averaging, and linearization; and

- extraction of transfer functions.

The first few lines of code take the parameters values. This is quite useful and is recommended since you can run the code easily for different values of components.

Converters dynamical equation (variables M1, M2, vO1, M3, M4, and vO2 in the code) must be extracted manually using KVL and KCL. MATLAB®can do symbolic computation using the "syms" command. For more information on the command, type "help syms" in MATLAB's command line.

Variables MA1, MA2, and vO in the code are averaged variables (see Eq. (3.19) and (3.24)). As said before, the steady state currents and voltages can be found by taking the derivative terms equal to zero (see Eq. (3.20)). IL and VC are steady state values of the inductor current and capacitor voltage, respectively. This gives us the operating point of the converter.

We take the derivative at the operating point to find out the corresponding state space coefficient. The state space system has been converted into the transfer function form using "tf" command.

The program can be used to extract Boost or Buck-Boost dynamical equations as well. Only dynamical equations (variables M1, M2, vO1, M3, M4, and vO2 in the code) must be updated.

3.9 DYNAMIC OF A CCM CUK CONVERTER

We study the dynamics of a Ćuk converter operating in CCM to give further example. Schematic of a Ćuk converter is shown in Fig. 3.11. Output of a Ćuk converter can be either larger or smaller than that of input, and there is a polarity reversal on the output. Steady-state output voltage of an ideal (i.e., converter with 100% efficiency) CCM Cuk converter can be calculated with the aid of following equation:

$$V_O = -\frac{D}{1-D} V_{IN},$$ (3.42)

where D is the duty ratio of MOSFET gate signal.

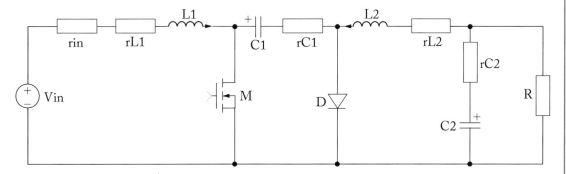

Figure 3.11: Schematic of a Ćuk converter.

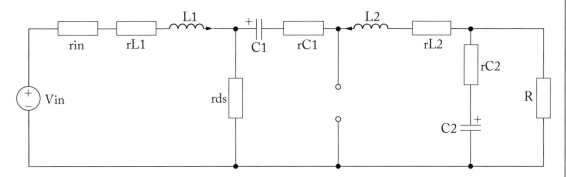

Figure 3.12: Equivalent circuit of a Cuk converter with closed MOSFET.

When MOSFET is closed, the diode is off. Equivalent circuit of this case is shown in Fig. 3.12. When MOSFET is closed (Fig. 3.12), using KVL and KCL, circuit dynamical equations are obtained as:

$$
\begin{cases}
\dfrac{d}{dt} i_{L_1} = -\dfrac{1}{L_1}\left(r_{in} + r_{L1} + r_{ds}\right) i_{L1} - \dfrac{r_{ds}}{L_1} i_{L_2} + \dfrac{1}{L_1} V_{IN} \\[2ex]
\dfrac{d}{dt} i_{L_2} = -\dfrac{r_{ds}}{L_2} i_{L1} - \dfrac{r_{C_1} + r_{L_2} + r_{ds} - \dfrac{r_{C_2} R}{r_{C_2} + R}}{L_2} i_{L_2} + \dfrac{1}{L_2} v_{C_1} + \dfrac{R}{(R + r_{C_2}) L_2} v_{C_2} \\[2ex]
\dfrac{d}{dt} v_{C_1} = -\dfrac{1}{C_1} i_{L_2} \\[2ex]
\dfrac{d}{dt} v_{C_2} = -\dfrac{R}{(R + r_{C_2}) C_2} i_{L_2} - \dfrac{1}{(R + r_{C_2}) C_2} v_{C_2}
\end{cases}
$$

$$
v_R = R\left(-i_{L_2} - C_2 \frac{dv_{C_2}}{dt}\right) = R\left(-i_{L_2} + \frac{R}{(R + r_{C_2})} i_{L_2} + \frac{1}{(R + r_{C_2})} v_{C_2}\right)
$$

$$
= R\left(\frac{-r_{C_2}}{(R + r_{C_2})} i_{L_2} + \frac{1}{(R + r_{C_2})} v_{C_2}\right). \tag{3.43}
$$

When MOSFET is opened, the diode is forward biased. An equivalent circuit of this case is shown in Fig. 3.13.

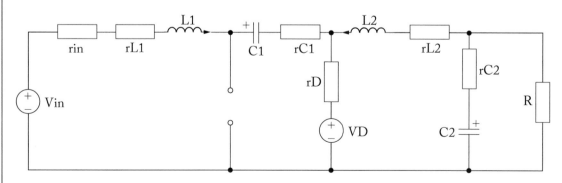

Figure 3.13: Equivalent circuit of a Ćuk converter with open MOSFET.

When MOSFET is opened (Fig. 3.13), using KVL and KCL, circuit dynamical equations are obtained as:

$$\begin{cases} \dfrac{d}{dt}i_{L_1} = -\dfrac{1}{L_1}\left(r_D + r_{in} + r_{L1} + r_{C_1}\right)i_{L1} - \dfrac{r_D}{L_1}i_{L2} - \dfrac{1}{L_1}v_{C_1} + \dfrac{1}{L_1}V_{IN} - \dfrac{1}{L_1}V_D \\[4mm] \dfrac{d}{dt}i_{L_2} = -\dfrac{r_D}{L_2}i_{L1} - \dfrac{r_D + r_{L_2} + \dfrac{Rr_{C_2}}{R + r_{C_2}}}{L_2}i_{L2} + \dfrac{R}{(R + r_{C_2})L_2}v_{C_2} - \dfrac{1}{L_2}V_D \\[4mm] \dfrac{d}{dt}v_{C_1} = \dfrac{1}{C_1}i_{L_1} \\[4mm] \dfrac{d}{dt}v_{C_2} = -\dfrac{R}{(R + r_{C_2})C_2}i_{L2} - \dfrac{1}{(R + r_{C_2})C_2}v_{C_2} \end{cases}$$

$$v_R = R\left(-i_{L_2} - C_2\dfrac{dv_{C_2}}{dt}\right) = R\left(-i_{L_2} + \dfrac{R}{(R + r_{C_2})}i_{L_2} + \dfrac{1}{(R + r_{C_2})}v_{C_2}\right)$$

$$= R\left(\dfrac{-r_{C_2}}{(R + r_{C_2})}i_{L_2} + \dfrac{1}{(R + r_{C_2})}v_{C_2}\right). \tag{3.44}$$

Assume a Ćuk converter with the following parameters: $R = 8\ \Omega$, $V_{IN} = 12$ V, $r_{in} = 0.5\ \Omega$, $L_1 = 700\ \mu H$, $r_{L_1} = 0.3\ \Omega$, $L_2 = 700\ \mu H$, $r_{L_2} = 0.3\ \Omega$, $C_1 = 22\ \mu F$, $r_{C_1} = 0.05\ \Omega$, $C_2 = 10\ \mu F$, $r_{C_2} = 0.05\ \Omega$, $r_D = 0.01\ \Omega$, $V_D = 0.7$ V, and $r_{ds} = 0.01\ \Omega$.

The previous program can be modified to obtain the small signal transfer functions of Cuk converter as follows. Only equations must be updated; the logic behind the program is the same as the Buck case. Output of program is shown in Fig. 3.14.

```
%Cuk converters dynamics
clc
clear all;
% Circuit Parameters
R=8;

VIN=12;
rin=.5;

L1=700e-6;
rL1=.3;

L2=700e-6;
rL2=.3;
```

```
C1=22e-6;
rC1=.05;

C2=10e-6;
rC2=.05;

rD=.01;
VD=.7;

rds=.1;

D=.6;
% Symbolic variables
syms iL1 iL2 vC1 vC2 vin vD d

%CLOSED MOSFET EQUATIONS
M1=(-(rin+rL1+rds)*iL1-rds*iL2+vin)/L1;
M2=(-rds*iL1-(rC1+rL2+rds-(rC2*R/(rC2+R)))*iL2+vC1+
        R/(rC2+R)*vC2)/L2;
M3=(-1*iL2)/C1;
M4=(-R/(R+rC2)*iL2-1/(R+rC2)*vC2)/C2;

vO1=-R*rC2/(R+rC2)*iL2+R/(R+rC2)*vC2;
%OPENED MOSFET EQUATIONS
M5=(-(rin+rL1+rC1+rD)*iL1-rD*iL2-vC1+vin-vD)/L1;
M6=(-rD*iL1-(rL2+rD+(R*rC2/(R+rC2)))*iL2+(R/(R+rC2))*vC2-vD)/L2;
M7=1*iL1/C1;
M8=(-1/(R+rC2)*vC2-R/(R+rC2)*iL2)/C2;

vO2=-R*rC2/(R+rC2)*iL2+R/(R+rC2)*vC2;
%AVERAGING
MA1= simplify(d*M1+(1-d)*M5);
MA2= simplify(d*M2+(1-d)*M6);
MA3= simplify(d*M3+(1-d)*M7);
MA4= simplify(d*M4+(1-d)*M8);

vO=d*vO1+(1-d)*vO2;
% DC OPERATING POINT CALCULATION
```

```
MA_DC_1=subs(MA1,[vin vD d],[VIN VD D]);
MA_DC_2=subs(MA2,[vin vD d],[VIN VD D]);
MA_DC_3=subs(MA3,[vin vD d],[VIN VD D]);
MA_DC_4=subs(MA4,[vin vD d],[VIN VD D]);


DC_SOL=solve(MA_DC_1==0,MA_DC_2==0,MA_DC_3==0,MA_DC_4==0,
            'iL1','iL2','vC1','vC2');


IL1=eval(DC_SOL.iL1);
IL2=eval(DC_SOL.iL2);
VC1=eval(DC_SOL.vC1);
VC2=eval(DC_SOL.vC2);

%LINEARIZATION
% x=[iL1;iL2;vC1;vC2]
%u=[vin;d] where d=duty and
A11=subs(simplify(diff(MA1,iL1)),[iL1 iL2 vC1 vC2 d vD],
        [IL1 IL1 VC1 VC2 D VD]);
A12=subs(simplify(diff(MA1,iL2)),[iL1 iL2 vC1 vC2 d vD],
        [IL1 IL1 VC1 VC2 D VD]);
A13=subs(simplify(diff(MA1,vC1)),[iL1 iL2 vC1 vC2 d vD],
        [IL1 IL1 VC1 VC2 D VD]);
A14=subs(simplify(diff(MA1,vC2)),[iL1 iL2 vC1 vC2 d vD],
        [IL1 IL1 VC1 VC2 D VD]);


A21=subs(simplify(diff(MA2,iL1)),[iL1 iL2 vC1 vC2 d vD],
        [IL1 IL1 VC1 VC2 D VD]);
A22=subs(simplify(diff(MA2,iL2)),[iL1 iL2 vC1 vC2 d vD],
        [IL1 IL1 VC1 VC2 D VD]);
A23=subs(simplify(diff(MA2,vC1)),[iL1 iL2 vC1 vC2 d vD],
        [IL1 IL1 VC1 VC2 D VD]);
A24=subs(simplify(diff(MA2,vC2)),[iL1 iL2 vC1 vC2 d vD],
        [IL1 IL1 VC1 VC2 D VD]);


A31=subs(simplify(diff(MA3,iL1)),[iL1 iL2 vC1 vC2 d vD],
        [IL1 IL1 VC1 VC2 D VD]);
A32=subs(simplify(diff(MA3,iL2)),[iL1 iL2 vC1 vC2 d vD],
        [IL1 IL1 VC1 VC2 D VD]);
A33=subs(simplify(diff(MA3,vC1)),[iL1 iL2 vC1 vC2 d vD],
```

```
                [IL1 IL1 VC1 VC2 D VD]);
A34=subs(simplify(diff(MA3,vC2)),[iL1 iL2 vC1 vC2 d vD],
            [IL1 IL1 VC1 VC2 D VD]);

A41=subs(simplify(diff(MA4,iL1)),[iL1 iL2 vC1 vC2 d vD],
            [IL1 IL1 VC1 VC2 D VD]);
A42=subs(simplify(diff(MA4,iL2)),[iL1 iL2 vC1 vC2 d vD],
            [IL1 IL1 VC1 VC2 D VD]);
A43=subs(simplify(diff(MA4,vC1)),[iL1 iL2 vC1 vC2 d vD],
            [IL1 IL1 VC1 VC2 D VD]);
A44=subs(simplify(diff(MA4,vC2)),[iL1 iL2 vC1 vC2 d vD],
            [IL1 IL1 VC1 VC2 D VD]);

A=eval([A11 A12 A13 A14;
        A21 A22 A23 A24;
        A31 A32 A33 A34;
        A41 A42 A43 A44]);

B11=subs(simplify(diff(MA1,vin)),[iL1 iL2 vC1 vC2 d vD],
            [IL1 IL1 VC1 VC2 D VD]);
B12=subs(simplify(diff(MA1,d)),[iL1 iL2 vC1 vC2 d vD],
            [IL1 IL1 VC1 VC2 D VD]);

B21=subs(simplify(diff(MA2,vin)),[iL1 iL2 vC1 vC2 d vD],
            [IL1 IL1 VC1 VC2 D VD]);
B22=subs(simplify(diff(MA2,d)),[iL1 iL2 vC1 vC2 d vD],
            [IL1 IL1 VC1 VC2 D VD]);

B31=subs(simplify(diff(MA3,vin)),[iL1 iL2 vC1 vC2 d vD],
            [IL1 IL1 VC1 VC2 D VD]);
B32=subs(simplify(diff(MA3,d)),[iL1 iL2 vC1 vC2 d vD],
            [IL1 IL1 VC1 VC2 D VD]);

B41=subs(simplify(diff(MA4,vin)),[iL1 iL2 vC1 vC2 d vD],
            [IL1 IL1 VC1 VC2 D VD]);
B42=subs(simplify(diff(MA4,d)),[iL1 iL2 vC1 vC2 d vD],
            [IL1 IL1 VC1 VC2 D VD]);

B=eval([B11 B12;
```

```
           B21 B22;
           B31 B32;
           B41 B42]);

CC1=subs(simplify(diff(v0,iL1)),[iL1 iL2 vC1 vC2 d vD],
         [IL1 IL1 VC1 VC2 D VD]);
CC2=subs(simplify(diff(v0,iL2)),[iL1 iL2 vC1 vC2 d vD],
         [IL1 IL1 VC1 VC2 D VD]);
CC3=subs(simplify(diff(v0,vC1)),[iL1 iL2 vC1 vC2 d vD],
         [IL1 IL1 VC1 VC2 D VD]);
CC4=subs(simplify(diff(v0,vC2)),[iL1 iL2 vC1 vC2 d vD],
         [IL1 IL1 VC1 VC2 D VD]);
CC=eval([CC1 CC2 CC3 CC4]);

H=tf(ss(A,B,CC,0));
vR_vin=H(1,1)
vR_d=H(1,2)
```

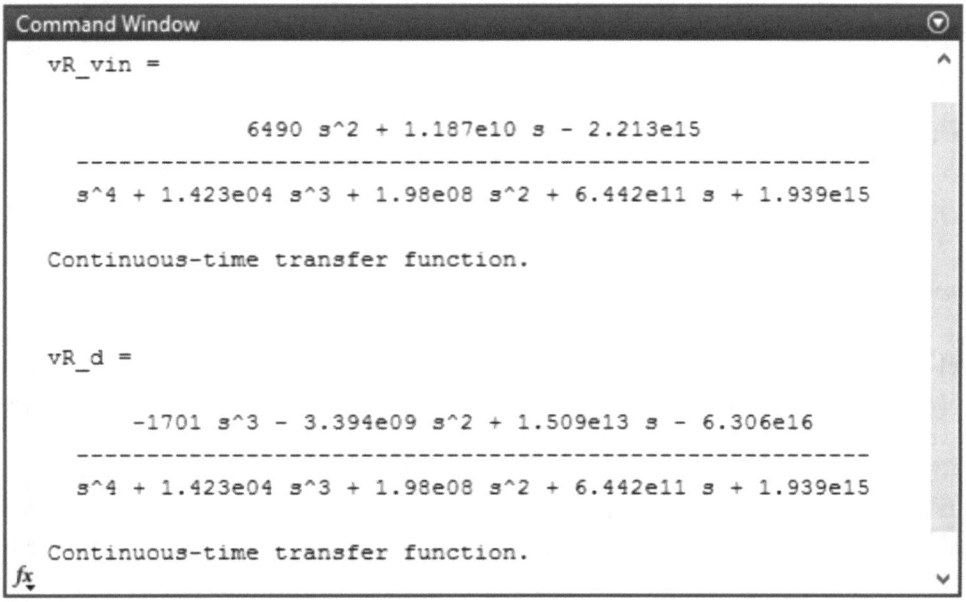

```
Command Window                                                    ⊙

  vR_vin =

                  6490 s^2 + 1.187e10 s - 2.213e15
     ------------------------------------------------------------
      s^4 + 1.423e04 s^3 + 1.98e08 s^2 + 6.442e11 s + 1.939e15

  Continuous-time transfer function.


  vR_d =

          -1701 s^3 - 3.394e09 s^2 + 1.509e13 s - 6.306e16
     ------------------------------------------------------------
      s^4 + 1.423e04 s^3 + 1.98e08 s^2 + 6.442e11 s + 1.939e15

  Continuous-time transfer function.
fx
```

Figure 3.14: Ćuk converter transfer functions.

Figures 3.15 and 3.16 show the control-to-output $\left(\frac{\tilde{v}_o(s)}{\tilde{d}(s)}\right)$ and the transfer function between input source and load voltage $\left(\frac{\tilde{v}_o(s)}{\tilde{v}_{in}(s)}\right)$.

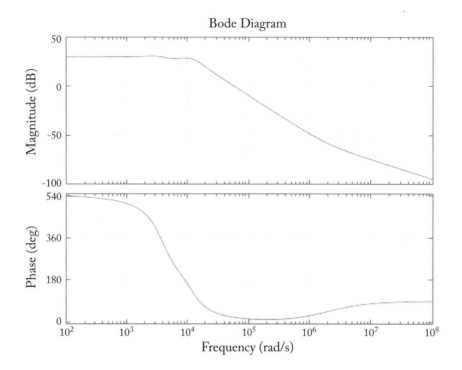

Figure 3.15: Bode plot of vO_d $= \frac{\tilde{v}_o(s)}{\tilde{d}(s)}$ for Ćuk converter.

3.10 KUCA: A SOFTWARE TOOL TO EXTRACT THE TRANSFER FUNCTIONS AUTOMATICALLY

A MATLAB®toolbox developed at the Kocaeli University's Power Electronic Research Group[2] can extract the small signal transfer function automatically. Figure 3.17 shows the main window of a developed toolbox.

The user enters the values of parameter and the transfer function he needs. Software extracts the equation automatically. Figure 3.18 shows the SEPIC converter analysis section of Kocaeli University Converter Analysis (KUCA) suite. After analysis either the algebraic transfer function or the Bode diagram is given.

[2]See Asadi, F. and Abut N. (2016). KUCA: Kocaeli University Converter Analysis simulation software in power electronics. *International Journal of Advanced and Applied Sciences*, 3(12):55–61.

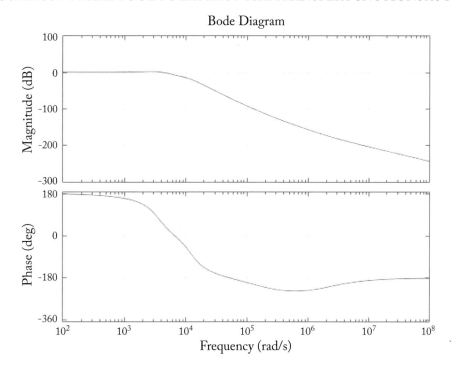

Figure 3.16: Bode plot of vO_vin $= \frac{\tilde{v}_o(s)}{\tilde{v}_{in}(s)}$ for Ćuk converter.

Figure 3.17: Main window of KUCA.

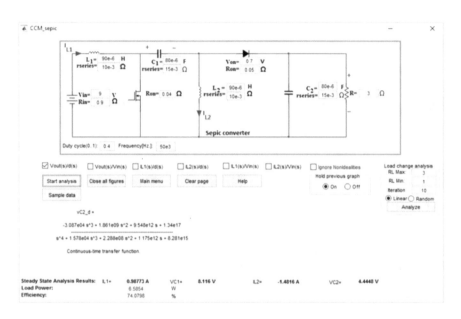

Figure 3.18: SEPIC converter analysis section of KUCA.

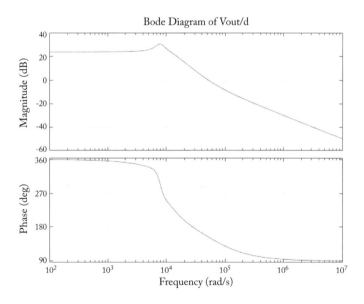

Figure 3.19: Bode diagram of the SEPIC converter (see Fig. 3.18).

3.11 OBTAINING THE BODE PLOT USING PSIM®

PSIM®can be used to extract the frequency response of converters. We draw the Buck converter schematic in PSIM®(Fig. 3.20). The required blocks can be found in the toolbar shown
in Fig. 3.21. The label block can be added to the simulation file using the icon shown in Fig. 3.22.
There is a clock icon available in the simulation file. It is called the "Simulation Control" block
and can be found in the "Simulate" menu. You can use the "Library Browser" icon (Fig. 3.23)
to search the PSIM®library of components to find the desired part. After clicking the "Library
Browser" icon, you enter its name in the opened dialog box as shown in Fig. 3.24. Used components settings are shown in Figs. 3.25–3.39. You can simulate the circuit either by pressing F8 or
clicking the run simulation icon shown in Fig. 3.40. The simulation result is shown in Fig. 3.41.
The icon ⊤ in Fig. 3.41 can be used to trace the waveform.

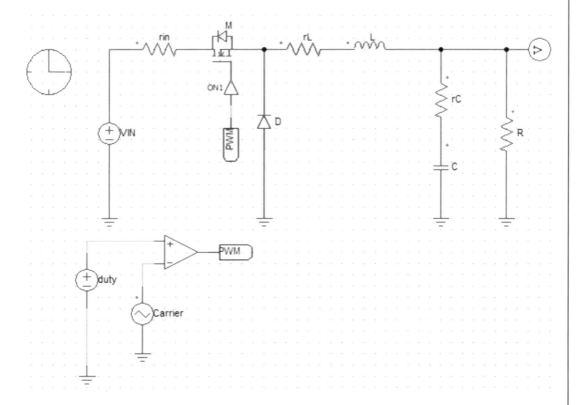

Figure 3.20: Simulation diagram of a Buck converter in PSIM®.

Figure 3.21: PSIM's elements.

Figure 3.22: Label element.

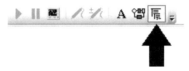

Figure 3.23: "Library Browser" icon.

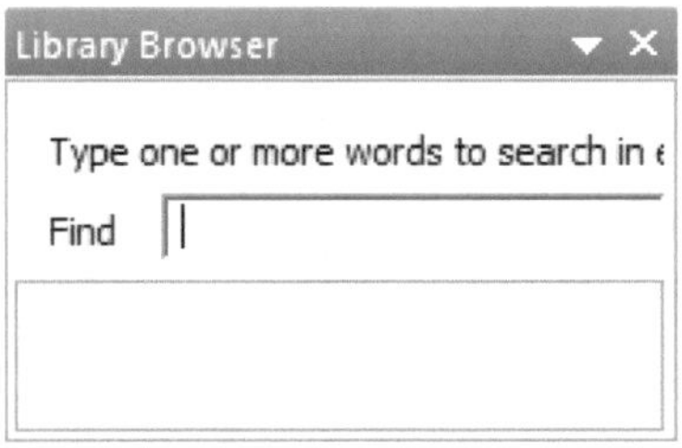

Figure 3.24: "Library Browser" window.

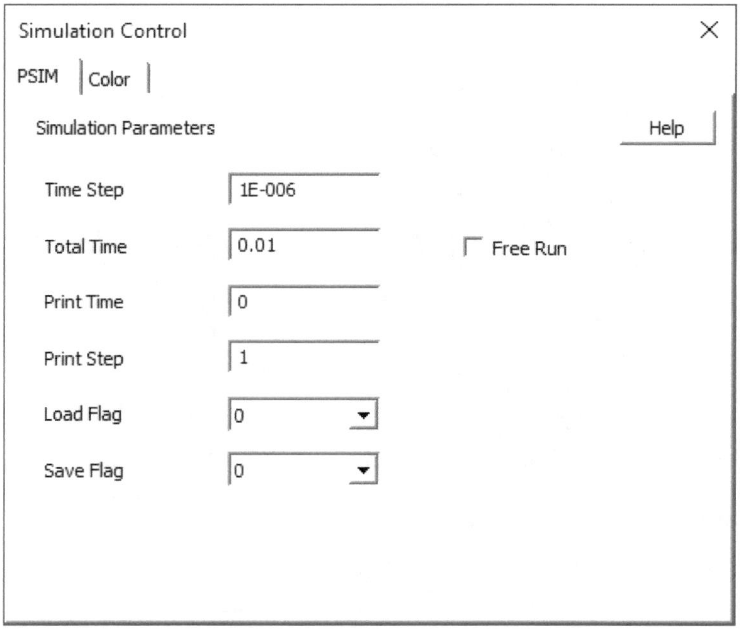

Figure 3.25: "Simulation Control" window.

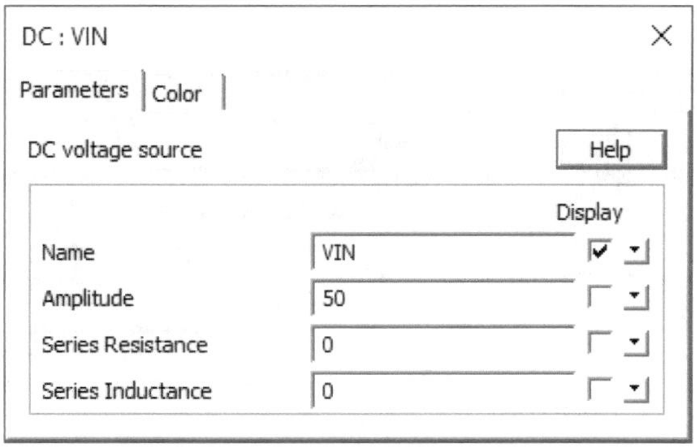

Figure 3.26: Voltage source "VIN" settings.

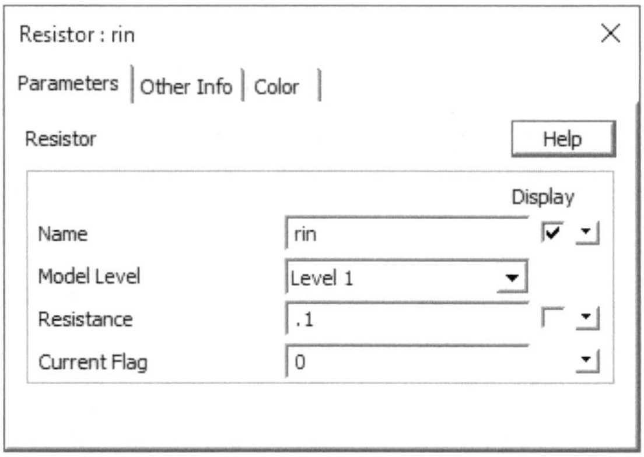

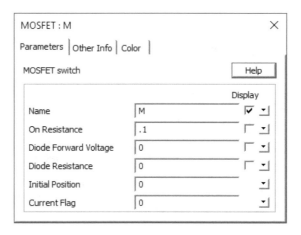

Figure 3.27: Resistor "rin" settings.

Figure 3.28: MOSFET "M" settings.

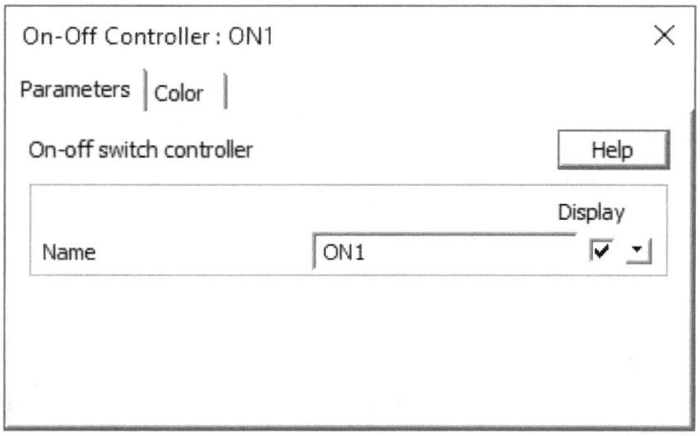

Figure 3.29: On-Off Controller "ON1" settings.

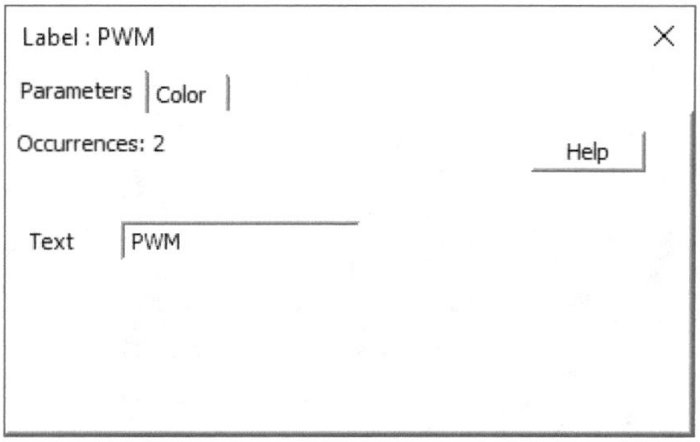

Figure 3.30: Label "PWM" settings.

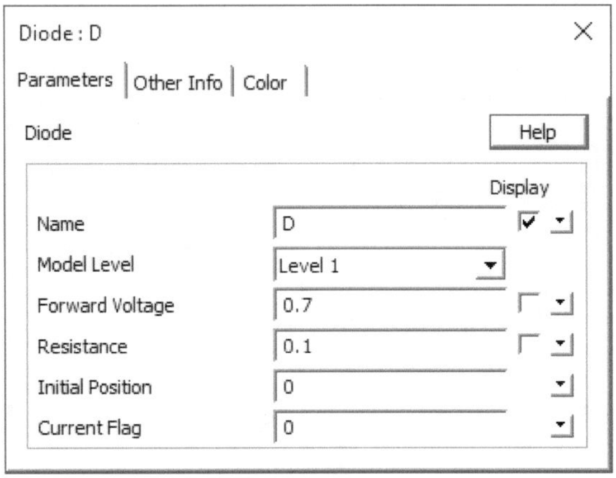

Figure 3.31: Diode "D" settings.

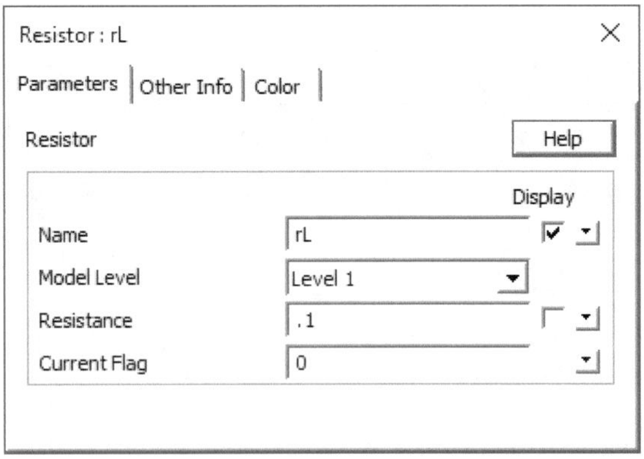

Figure 3.32: Resistor "rL" settings.

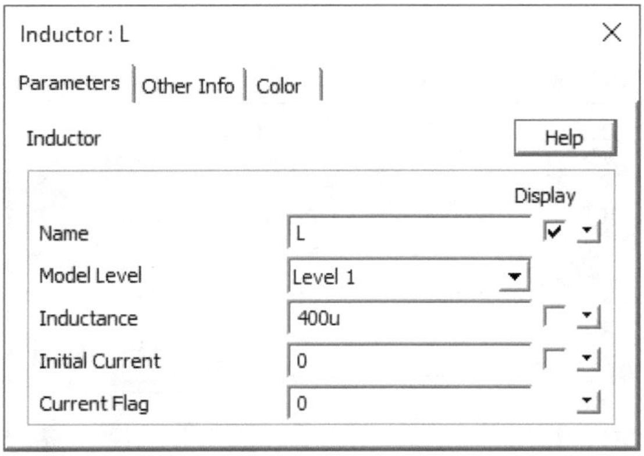

Figure 3.33: Inductor "L" settings.

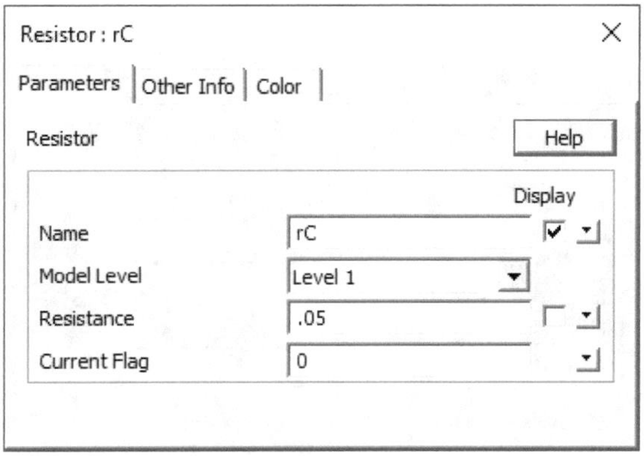

Figure 3.34: Resistor "rC" settings. rC plays the role of capacitor ESR.

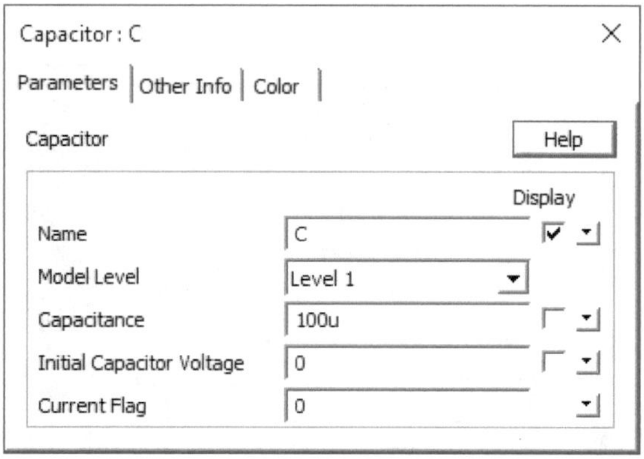

Figure 3.35: Capacitor "C" settings.

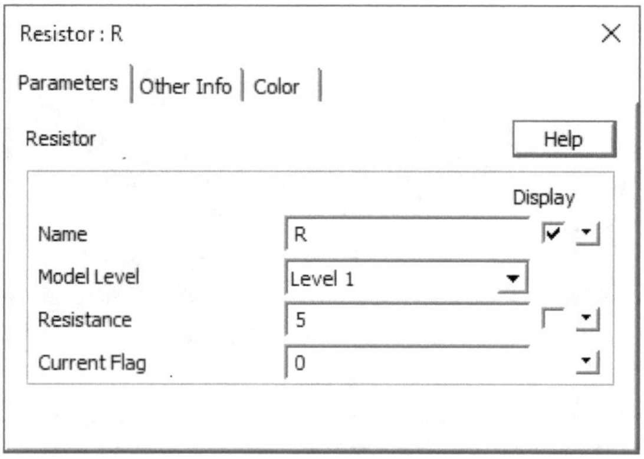

Figure 3.36: Resistor "R" settings.

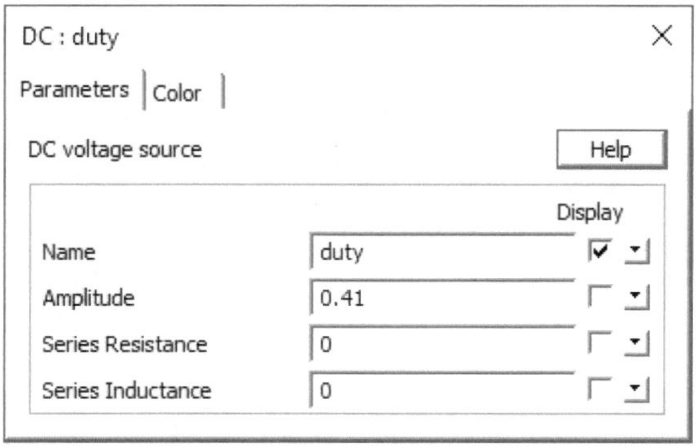

Figure 3.37: Votage source "duty" settings.

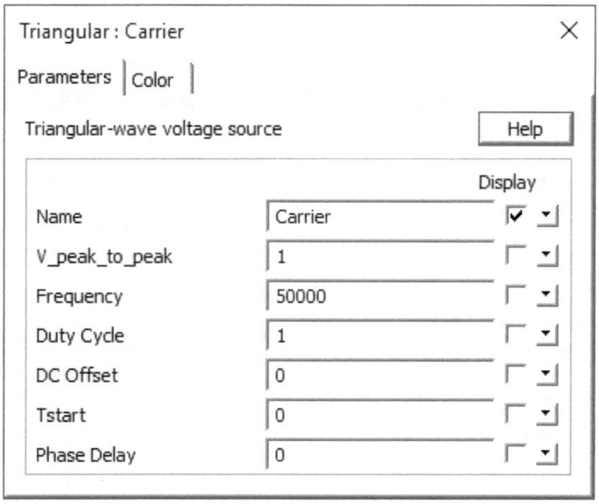

Figure 3.38: Triangular voltage source "Carrier" settings.

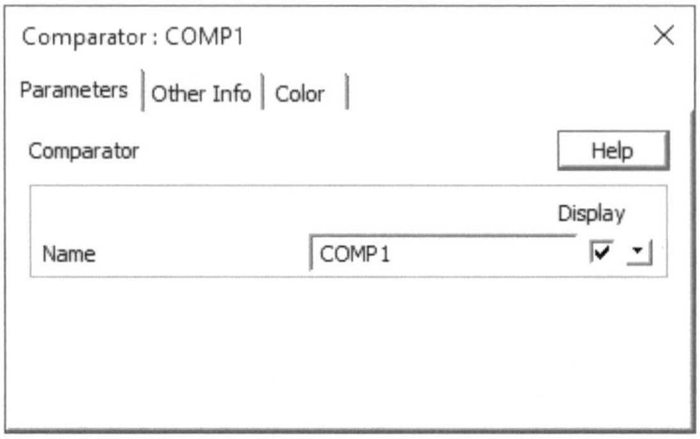

Figure 3.39: Comparator "COMP1" settings.

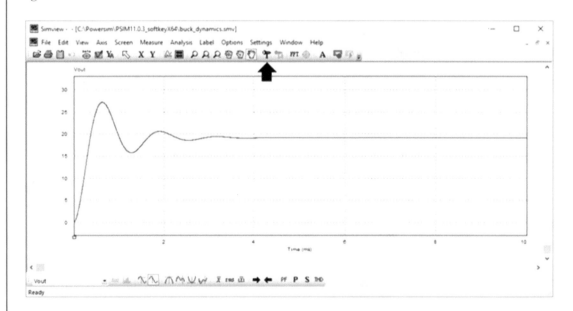

Figure 3.40: Run simulation icon.

Figure 3.41: Simulation result.

In order to obtain the small signal transfer functions, we change the simulation diagram to the one shown in Fig. 3.42. The block named V1 is a sinusoidal source which plays the role of perturbation (like $\tilde{d}$ in $d = D + \tilde{d}$). An "ac probe" and "AC Sweep" block is added to the simulation file as well. These blocks can be found using the "Library Browser," as shown in Fig. 3.43.

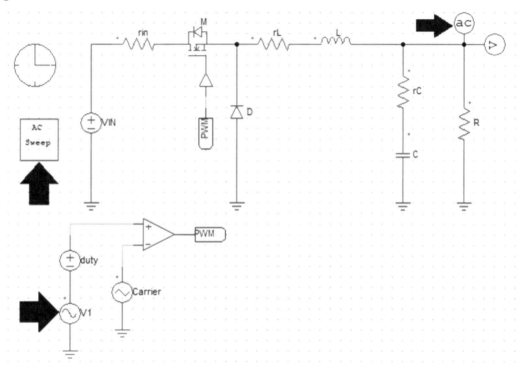

Figure 3.42: Adding the "AC Sweep" and "AC probe" block to the simulation file. We want to obtain the $\frac{\tilde{v}_o(s)}{\tilde{d}(s)}$ so we must place the sinusoidal perturbation block in series with duty.

There is no need to change the "V1" settings. You can keep the default values (Fig. 3.44). The "V1" settings have no effect on the "AC Sweep" analysis results.

We set the "AC Sweep" block as shown in Fig. 3.45. We set the "Start Amplitude" and "End Amplitude" to be quite small with respect to the operating point. In this case, the operating point is set by the block named "duty" (see Fig. 3.42) which has the value of 0.41. So, values which are less than $\frac{0.41}{10} = .041$ are suitable for "Start Amplitude" and "End Amplitude." The "End Amplitude" box gives you a chance to increase the perturbation as the simulation progress from lower frequencies toward higher frequencies. This leads to a smoother graph in comparison with a sweep done with a constant frequency independent perturbation specially in a high region of plot. So, it is recommended to fill the "End Amplitude" with a value greater than "Start Amplitude."

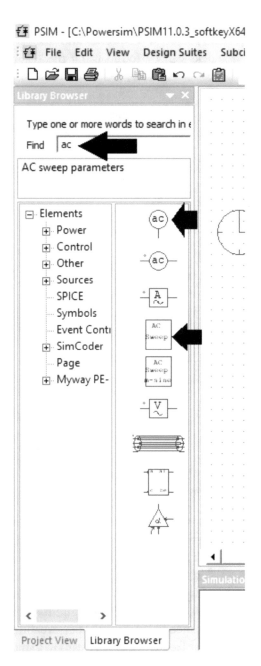

Figure 3.43: "AC probe" and "AC Sweep" blocks.

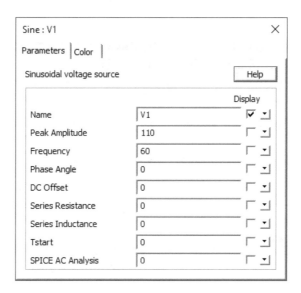

Figure 3.44: Sinusoidal voltage source "V1" settings.

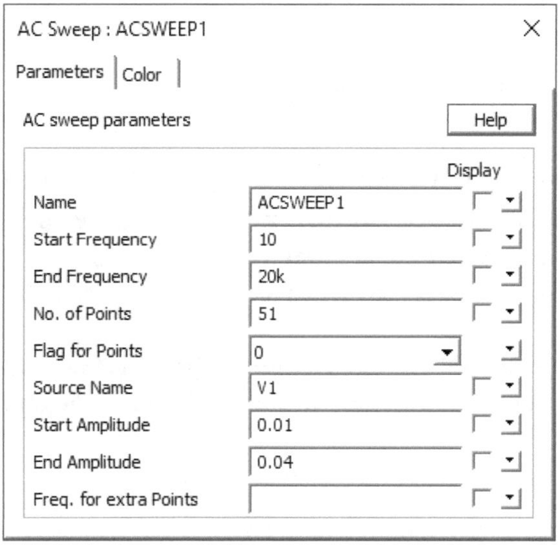

Figure 3.45: "AC Sweep" block settings.

You must enter the name of the perturbation source into the "Source Name" box. In this case, we used a block named "V1" as the perturbation source so we enter "V1" into the "Source Name" box. Half of the switching frequency is a good choice for "End Frequency."

Simulation result is shown in Fig. 3.46. You can use the icon ⌖ to read the graph values. The obtained result is compatible with that shown in Fig. 3.8.

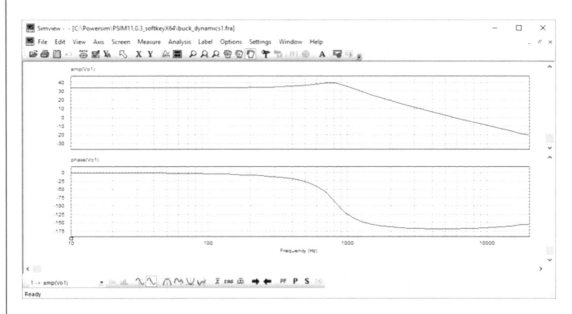

Figure 3.46: Bode plot of $\frac{\tilde{v}_o(s)}{\tilde{d}(s)}$.

You can obtain the $\frac{\tilde{v}_o(s)}{\tilde{d}(s)}$ as well if you place the perturbation block (V1) in series with the input source. In this case, V1 must be less than VIN. VIN is 50 V so values less than $\frac{50}{10} = 5$ V are good for "Start Amplitude" and "End Amplitude." We fill the "AC Sweep" block as shown in Fig. 3.48. The simulation result is shown in Fig. 3.49. The obtained result is compatible with that shown in Fig. 3.9.

As you may notice, the PSIM®cannot provide the algebraic transfer function. Only a numeric Bode plot is obtainable. A method to extract the algebraic transfer function is studied below.

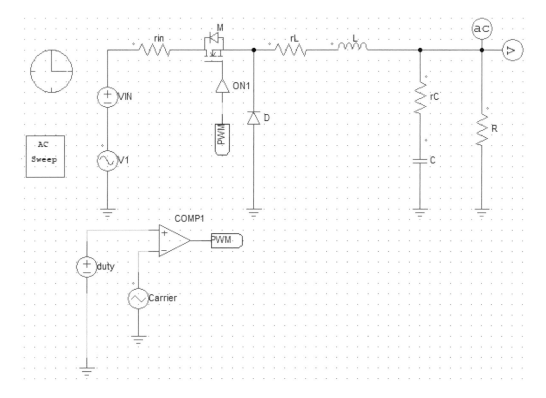

Figure 3.47: We want to obtain the $\frac{\tilde{v}_o(s)}{\tilde{v}_{in}(s)}$ so we must place the sinusoidal perturbation block in series with the input voltage source.

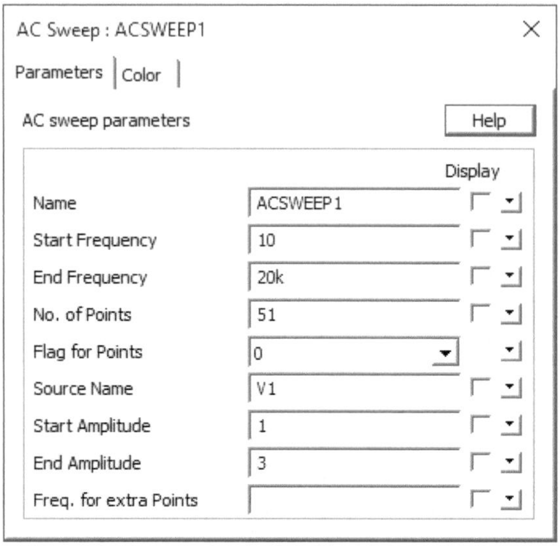

Figure 3.48: "AC Sweep" block (used in Fig. 3.47) settings.

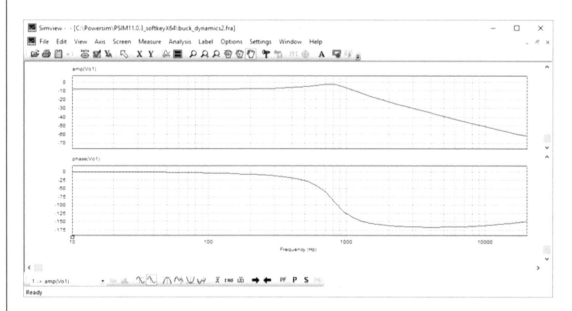

Figure 3.49: Bode plot of $\frac{\tilde{v}_o(s)}{\tilde{v}_{in}(s)}$.

3.12 OBTAINING THE ALGEBRAIC TRANSFER FUNCTION OF CONVERTERS WORKING IN DCM

Dynamics of converters working in DCM have been studied in many papers and books.[3] We introduce a method to obtain the algebraic transfer functions of converters using software tools. The method can be used with converters working in DCM or CCM. So, it is a general procedure.

In order to make comparison possible, we use an example which has been studied before and the answer is known. Assume a Buck converter studied before. The Buck converter's parameters are given below (see Fig. 3.1):

$R = 5 \Omega$, $V_{in} = 50$ V, $r_{in} = 0.1 \Omega$, $L = 400 \mu$H, $r_L = 0.1 \Omega$, $C = 100 \mu$F, $r_C = 0.05 \Omega$, $D = 0.41$, $r_{ds} = 0.1 \Omega$, $r_D = 0.1 \Omega$, and $V_D = 0.7$ V.

The control-to-output transfer function $\left(\frac{\tilde{v}_o(s)}{\tilde{d}(s)} \right)$ has been calculated as:

$$\frac{\tilde{v}_o(s)}{\tilde{d}(s)} = \frac{6184s + 1.237 \times 10^9}{s^2 + 2574s + 2.568 \times 10^7}.$$

We draw the simulation diagram in PSIM®and do an "AC Sweep," as done in the previous section. After seeing the result (Fig. 3.46), we use the File menu (Fig. 3.50) to save the obtained result as a comma separated file (Fig. 3.51). We save it under the name "`freqdata.csv`," as shown in Fig. 3.51.

If we open the "`freqdata.csv`" in Notepad or Microsoft Excel® we see a three-column file (Fig. 3.52). The first column is the frequency in Hz, the second column is the magnitude in dB, and the third column is the angle in degrees. For example, according to Fig. 3.52, the converter control-to-output transfer function has a magnitude of 33.6487 db and phase angle of $-0.374147°$ at 10 Hz (first row).

We enter the MATLAB®environment and use the command "csvread" to read the csv file produced by PSIM®. As shown in Fig. 3.53, an error has been generated. You can solve the problem easily by removing the first line of "`freqdata.csv`." You can open the "`freqdata.csv`" in Notepad and remove the first line, as shown in Fig. 3.54. After removing the first line we can enter the frequency response data to MATLAB®without any error, as shown in Fig. 3.55. We use the following codes to form a copy of what we have seen in PSIM®environment (i.e., Fig. 3.56).

We can see the transfer function of variable sys by writing the MATLAB®command "bode(sys), grid minor." The result is shown in Fig. 3.57.

---

[3]See Chapter 11: "AC and DC Equivalent Circuit Modeling of the Discontinuous Conduction Mode" in *Fundamental of Power Electronics* by Erickson and Maksimovic.

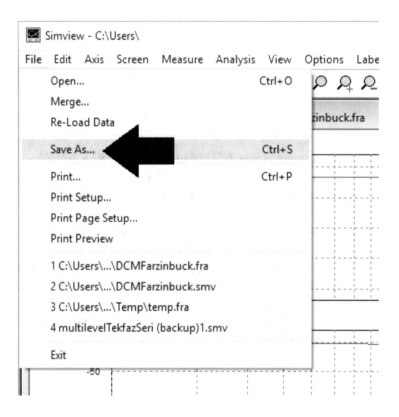

Figure 3.50: The obtained bode plot is saved by clicking the "Save As….."

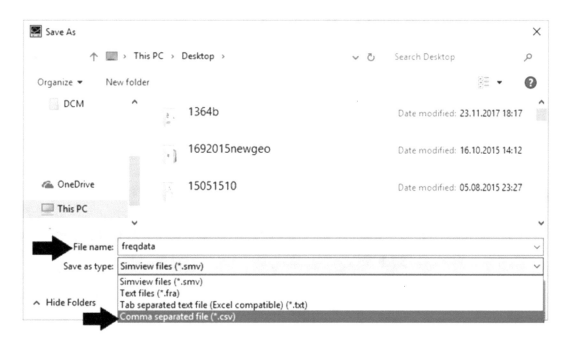

Figure 3.51: "Save as type: Comma separated file (∗.csv)" is used to produce the comma separated file.

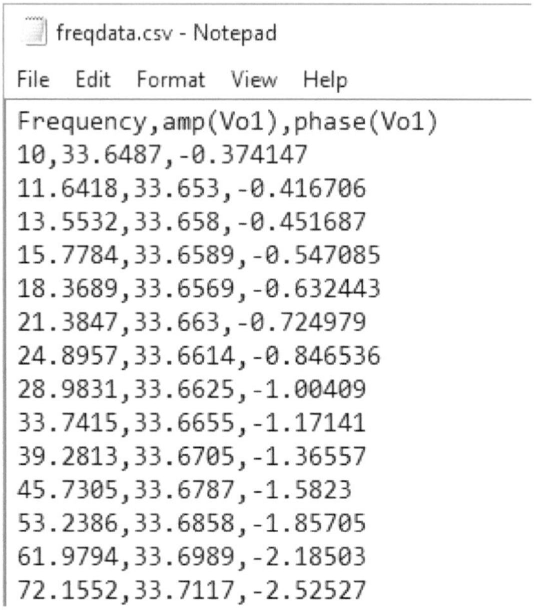

```
freqdata.csv - Notepad

File   Edit   Format   View   Help

Frequency,amp(Vo1),phase(Vo1)
10,33.6487,-0.374147
11.6418,33.653,-0.416706
13.5532,33.658,-0.451687
15.7784,33.6589,-0.547085
18.3689,33.6569,-0.632443
21.3847,33.663,-0.724979
24.8957,33.6614,-0.846536
28.9831,33.6625,-1.00409
33.7415,33.6655,-1.17141
39.2813,33.6705,-1.36557
45.7305,33.6787,-1.5823
53.2386,33.6858,-1.85705
61.9794,33.6989,-2.18503
72.1552,33.7117,-2.52527
```

Figure 3.52: Produced .csv file is opened in Notepad.

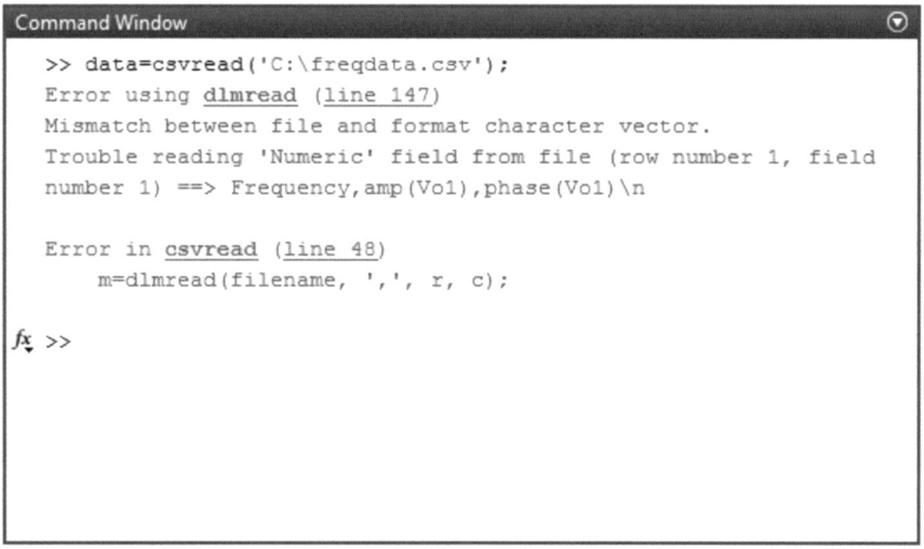

```
Command Window

>> data=csvread('C:\freqdata.csv');
Error using dlmread (line 147)
Mismatch between file and format character vector.
Trouble reading 'Numeric' field from file (row number 1, field
number 1) ==> Frequency,amp(Vo1),phase(Vo1)\n

Error in csvread (line 48)
    m=dlmread(filename, ',', r, c);

fx >>
```

Figure 3.53: Produced error message.

Figure 3.54: `freqdata.csv` file without first line.

Figure 3.55: Entering the `freqdata.csv` file into MATLAB®.

```
Command Window                                                              ⊙
   >> data=csvread('C:\freqdata.csv');
   Error using dlmread (line 147)
   Mismatch between file and format character vector.
   Trouble reading 'Numeric' field from file (row number 1, field
   number 1) ==> Frequency,amp(Vo1),phase(Vo1)\n

   Error in csvread (line 48)
       m=dlmread(filename, ',', r, c);

   >> data=csvread('C:\freqdata.csv');
   >> w=2*pi*data(:,1);% w is the frequency in rad/s
   >> val=10.^(data(:,2)/20).*exp(j*data(:,3)*pi/180);% frequency response values
   >> sys = frd(pf,w);
fx >>
```

Figure 3.56: Making the frequency response data (frd) of imported data.

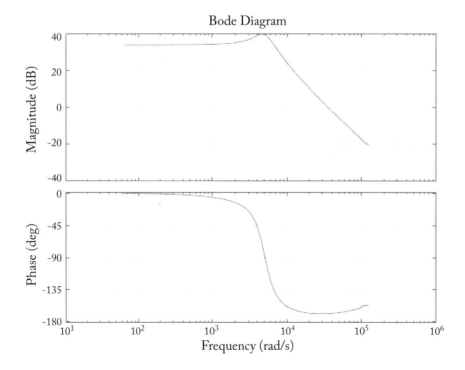

Figure 3.57: Result of "bode(sys)" command.

So, up to this point we have transferred the data from PSIM®to MATLAB®. We can use the "tfest" command to estimate a transfer function for the imported data. Results are shown in Fig. 3.58. If you compare the obtained transfer function with the result obtained before $\left(\frac{\tilde{v}_o(s)}{\tilde{d}(s)} = \frac{6184s+1.237\times10^9}{s^2+2574s+2.568\times10^7} \right)$ you will see the close similarity.

"Fit to estimation data" shows the goodness of fit. In this case, "Fit to estimation data" is about 100%. So, it says that the estimated model will overlap the imported data from PSIM®very well.

We can draw both the frequency response imported into MATLAB®(variable sys) and the estimated transfer function (variable H) on the same graph using the command "bode(sys,H), grid minor." The obtained result is shown in Fig. 3.59. As can be seen, both of them are overlapped which shows the goodness of estimated model.

With the aid of this method one can obtain an algebraic transfer function for the PSIM's Bode diagram. There is no limitation on the operation mode of the converter. So, the method can be applied to either CCM or DCM converters.

3.13 DYNAMICS OF PWM MODULATOR

The circuit shown in Fig. 3.60 shows the modulator required to transform the output of controller into pulses.

Assume that the "Carrier" signal is a ramp of amplitude V_{tri}, as shown in Fig. 3.61. Assume that a constant signal of V_{ref} is connected to the positive pin of the comparator. In this case, the output is like that shown in Fig. 3.61.

We want to obtain the duty ratio of produced pulse. The ramp carrier equation can be written as:

$$v_{carrier}(t) = \frac{V_{tri}}{T} \times t, \qquad 0 < t < T.$$

At $t = T_x$ the ramp carrier reaches the reference voltage connected to the positive pin of comparator. So, one can find the intersection by solving $v_{carrier}(T_x) = V_{ref}$

$$\frac{V_{tri}}{T} \times T_x = V_{ref} \implies T_x = \frac{V_{ref}}{V_{tri}} \times T.$$

The duty ratio of output pulse can be obtained as:

$$D = \frac{T_x}{T} = \frac{\frac{V_{ref}}{V_{tri}} \times T}{T} = \frac{V_{ref}}{V_{tri}} = \frac{1}{V_{tri}} \times V_{ref}.$$

So, the modulator looks like a simple gain[4] from the control view point, as shown in Fig. 3.62.

---

[4]There are other models which consider the delay in the PWM modulator as well. For the purpose of this book, simple gain model is enough.

```
Command Window                                                    ⊙

  >> data=csvread('C:\freqdata.csv');
  Error using dlmread (line 147)
  Mismatch between file and format character vector.
  Trouble reading 'Numeric' field from file (row number 1, field
  number 1) ==> Frequency,amp(Vo1),phase(Vo1)\n

  Error in csvread (line 48)
      m=dlmread(filename, ',', r, c);

  >> data=csvread('C:\freqdata.csv');
  >> w=2*pi*data(:,1);% w is the frequency in rad/s
  >> val=10.^(data(:,2)/20).*exp(j*data(:,3)*pi/180);% frequency response values
  >> sys = frd(val,w);
  >> bode(sys),grid minor
  >> bode(sys)
  >> H=tfest(sys,2)

  H =

       4328 s + 1.237e09
     -----------------------
     s^2 + 2576 s + 2.568e07

  Continuous-time identified transfer function.

  Parameterization:
     Number of poles: 2    Number of zeros: 1
     Number of free coefficients: 4
     Use "tfdata", "getpvec", "getcov" for parameters and their uncertainties.

  Status:
  Estimated using TFEST on frequency response data "sys".
  Fit to estimation data: 99.95%  ◀━━━
  FPE: 0.0003518, MSE: 0.0003252
fx >>
```

Figure 3.58: Result of "tfest(sys,2)" command. "tfest(sys,2)" estimate the best second-order transfer function to frd object "sys."

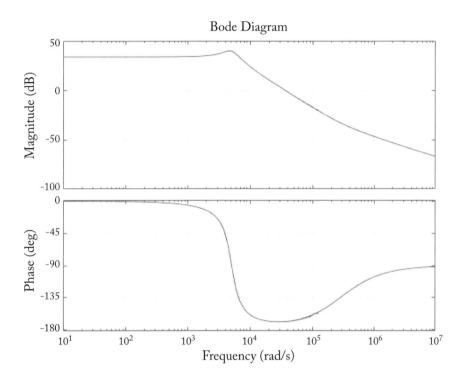

Figure 3.59: Estimated and imported data are drawn on the same axes.

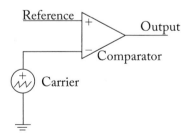

Figure 3.60: PWM generator.

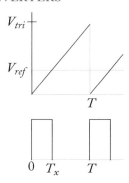

Figure 3.61: **PWM** generation process.

Figure 3.62: Small signal model of PWM generator shown in Fig. 3.60.

In this book we take the $V_{tri} = 1$ so the block reduces to $\frac{1}{1} = 1$. If you use other values of V_{tri} you must consider its effect. The block diagram shown in Fig. 3.63 shows the anatomy of a PWM DC-DC converter.

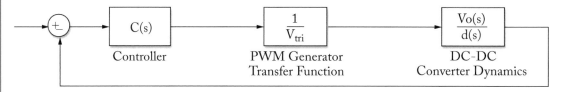

Figure 3.63: Feedback control of DC-DC converters.

As can be seen, the controller block (C(s)) sees two blocks in front of it $\left(\frac{1}{V_{tri}} \text{ and } \frac{\tilde{v}_{o(s)}}{\tilde{d}(s)} \right)$. So, the combination of these two blocks $\left(\frac{1}{V_{tri}} \times \frac{\tilde{v}_{o(s)}}{\tilde{d}(s)} \right)$ plays the role of plant. Assume the Buck converter we studied before. We obtained $\frac{\tilde{v}_o(s)}{\tilde{d}(s)} = \frac{6184s+1.237\times10^9}{s^2+2574s+2.568\times10^7}$. If you plan to use a ramp of amplitude 3 in the PWM generator, you must design the controller for $\frac{1}{3} \times \frac{6184s+1.237\times10^9}{s^2+2574s+2.568\times10^7}$ and if you use the ramp of amplitude 1, you must design the controller for $\frac{1}{1} \times \frac{6184s+1.237\times10^9}{s^2+2574s+2.568\times10^7} = \frac{6184s+1.237\times10^9}{s^2+2574s+2.568\times10^7}$.

We use a ramp signal of amplitude one in this book for the sake of simplicity. The "PWM Generator DC-DC" block in Simscape library Simulink®uses a ramp of amplitude 1 as well.

3.14 CONCLUSION

In this chapter we focused on the problem of modeling DC-DC converters. An important analytical tool named SSA has been studied. Computer methods of obtaining the converter transfer function is shown as well.

The models extracted in this chapter will be used in the next chapter to design controllers.

REFERENCES

[1] Daniel Hart, *Power Electronics*, McGraw Hill, 2011.

[2] Robert Erikson and Dragan Maksimovic, *Fundamentals of Power Electronics*, Springer, 2001. DOI: 10.1007/b100747.

[3] Simon Ang and Alejandro Oliva, *Power Switching Converters*, Taylor & Francis, 2005.

[4] Marian K. Kazimierczuck, *Pulse Width Modulated DC-DC Power Converters*, John Wiley, 2012. DOI: 10.1002/9780470694640.

[5] Seddik Bacha, Iulian Munteanu, and Antoneta Iluliana Bratcu, *Power Electronics Converters Modeling and Control*, Springer, 2014. DOI: 10.1007/978-1-4471-5478-5.

[6] Francesco Vasca and Luigi Iannelli, *Dynamics and Control of Switched Electronic Systems*, Springer, 2012. DOI: 10.1007/978-1-4471-2885-4.

[7] Teuvo Suntio, *Dynamic Profile of Switched Mode Converter: Modeling, Analysis and Control*, Wiley VCH, 2009. DOI: 10.1002/9783527626014.ch1.

[8] Jian Sun, D. M. Mitchell, M. F. Greuel, P. T. Krein, and R. M. Bass, Averaged modeling of PWM converters operating in discontinuous conduction mode, *IEEE Transactions on Power Electronics*, 16(4), pp. 482–492, July 2001. DOI: 10.1109/63.931052.

CHAPTER 4 ·

Controller Design

4.1 INTRODUCTION

The previous chapter introduced the modeling process in DC-DC converters. We use the extracted models to design controllers in this chapter.

Traditionally controllers were designed using pencil and paper. Nowadays, software is an integral part of control engineering. Based on this fact we try to show you how to solve the control design problem for DC-DC converters using software tools. We mainly use MATLAB®since it provides a design and simulation environment for the user.

We emphasize PID control in this chapter since it is simple and cheap. About 90% of real-world applications are controlled by PID controllers. This shows the importance of PID controllers. Some of the more advanced methods (like loop shaping) are introduced in this chapter for an interested reader.

We use the Buck converter of the previous chapter as an illustrative example in this chapter. The methods introduced in this chapter are general methods. You can apply them successfully to other types of DC-DC converters. We design the continuous time controllers in this chapter. You can obtain the discrete time version by using continuous to discrete transforms or "c2d" command in MATLAB®if required. References at the end of the chapter give more information on the subject. An interested reader is encouraged to study them.

4.2 CONTROLLER DESIGN FOR A BUCK CONVERTER

Assume a Buck Converter (Fig. 4.1) with the following parameters:
$R = 5\,\Omega, V_{in} = 50\,\text{V}, r_{in} = 0.1\,\Omega, L = 400\,\mu\text{H}, r_L = 0.1\,\Omega, C = 100\,\mu\text{F}, r_C = 0.05\,\Omega, D = 0.41, r_{ds} = 0.1\,\Omega, r_D = 0.1\,\Omega$, and $V_D = 0.7\,\text{V}$.

We obtained the small signal transfer functions as (see previous chapter):

$$\frac{\tilde{v}_o(s)}{\tilde{v}_{in}(s)} = \frac{50.74s + 1.015 \times 10^7}{s^2 + 2574s + 2.568 \times 10^7}$$
$$\frac{\tilde{v}_o(s)}{\tilde{d}(s)} = \frac{6184s + 1.237 \times 10^9}{s^2 + 2574s + 2.568 \times 10^7}.$$

One can draw the following block diagram for the Buck converter. The input voltage changes act like disturbance and the converter is controlled with the aid of duty ratio of input pulse.

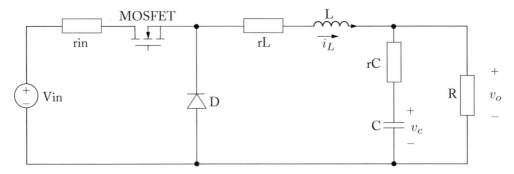

Figure 4.1: Buck converter.

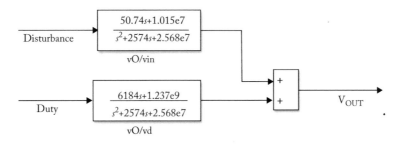

Figure 4.2: Transfer functions of a Buck converter.

One may add other disturbances like output load changes to this model. But it is enough for the purpose of illustration. We want to turn the open loop system into a closed loop one. We can use the structure shown in Fig. 4.3.

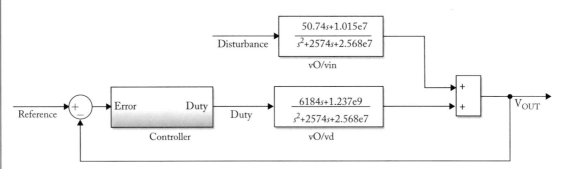

Figure 4.3: Closed-loop control of a Buck converter. Controller set the duty ratio.

In this block diagram, the controller senses the difference between the reference and based on the difference set the duty. This is called Voltage Mode Control (VMC) since the feedback is taken from output voltage. One may prefer to take feedback from the current instead of voltages. This type of control is called Current Mode Control (CMC).

We want to design the block named "Controller" in Fig. 4.3. Assume that we want to design a PI controller for this system. First of all, we must enter the transfer function for plant $\left(\frac{\tilde{v}_o(s)}{\tilde{d}(s)}\right)$ into the MATLAB®. This can be done with the aid of "tf" command.

```
Command Window                                        ⊙
    >> H=tf([6184 1.237e9],[1 2574 2.568e7]);
fx >>
```

Figure 4.4: Entering the transfer functions into MATLAB®with the aid of "tf" command.

We use the "pidTuner" command to tune the PI controller for the Buck converter.

```
Command Window                                        ⊙
    >> H=tf([6184 1.237e9],[1 2574 2.568e7]);
    >> pidTuner(H)
fx >>
```

Figure 4.5: The "pidTuner" command is used to tune the PID controller coefficients.

After pressing the Enter key, the window shown in Fig. 4.6 will open.

As shown in Fig. 4.7, values suggested by MATLAB®for PID controller is shown in the right lower corner of opened window. Kp, Ki, and Kd show the coefficient of proportional term, integral term, and derivative term, respectively.

First of all, you must choose your desired controller type. You can do this by clicking on the "Type" button, as shown in Fig. 4.8. We choose PI for this example.

You can use the "Form: Standard" (Fig. 4.9) to convert the proportional, integral, and derivative term coefficients into time constant format.

You can use the slider to obtain the desired time response. For example, the response shown in Fig. 4.12 seems quite good. Tracking error is zero and settling time is about 6 ms. We can use "Domain: Frequency" to see the bandwidth and phase margin. Generally, phase margin must be greater than 45°.

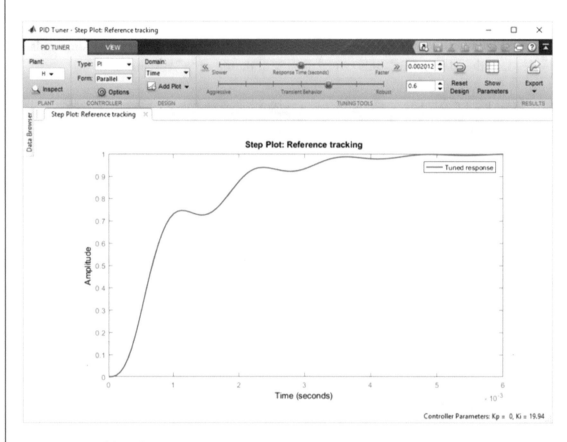

Figure 4.6: "pidTuner" application environment.

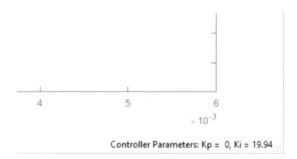

Figure 4.7: Kp and Ki show the PID controller proportional gain and integral gain, respectively.

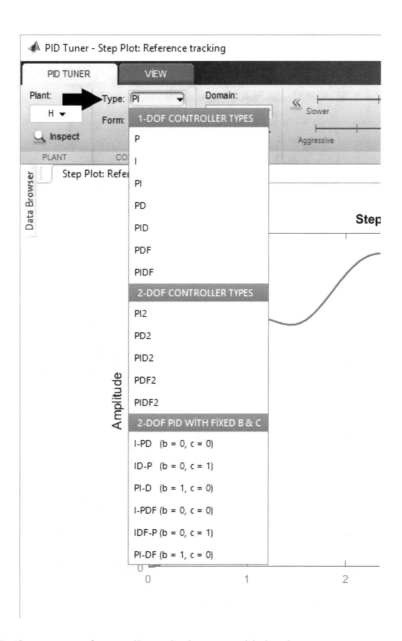

Figure 4.8: Different type of controllers which are tunable by the PID Tuner.

Figure 4.9: "Form" drop-down list.

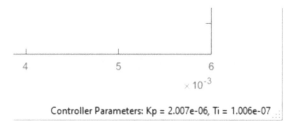

Figure 4.10: Controller parameters shown in time constant form.

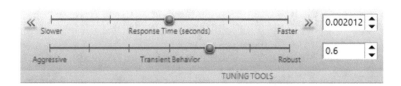

Figure 4.11: Sliders are used to obtain the desired response.

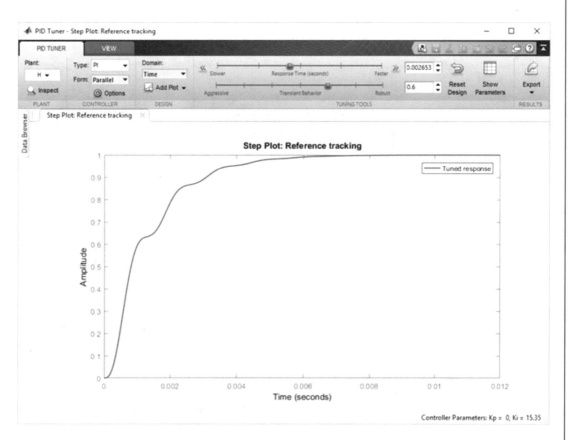

Figure 4.12: An example of a good response.

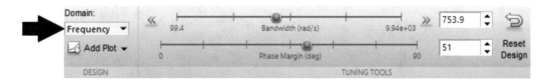

Figure 4.13: Tuning the controller in frequency domain.

One can complete the whole of tuning process in the frequency domain. When trial and errors are finished you can export the designed controller into MATLAB®workspace. To do this, we click the Export button. This pops up the window shown in Fig. 4.14. If you click the "OK" button in this window, designed PI controller is transferred into the workspace with the name of "C". If you write "C" in the command window and press the Enter key you will see the designed controller equation (Fig. 4.16).

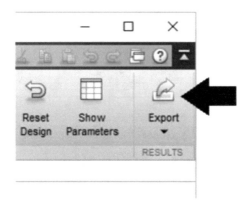

Figure 4.14: Transferring the designed controller into MATLAB®workspace.

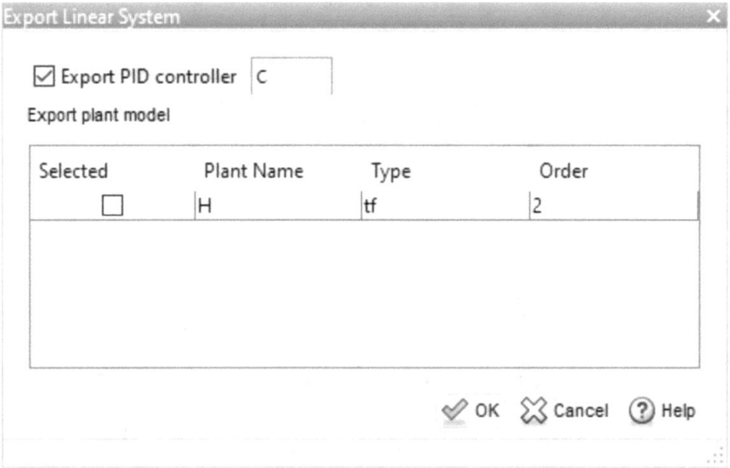

Figure 4.15: "Export Linear System" window.

```
Command Window                                              ⊙

   >> H=tf([6184 1.237e9],[1 2574 2.568e7]);
   >> pidTuner(H)
   >> C


   C =


           1
      Ki * ---
            s


      with Ki = 15.4


   Continuous-time I-only controller.

fx >>
```

Figure 4.16: Designed controller.

Obtaining a desired *output* response is not all the work. All the signals within the control loop must be within the allowed range. We can see the other signals in the loop simultaneously by adding them to the current window. This can be done with the aid of the "Add Plot" button shown in Fig. 4.17. Generally, "Control Effort" is added to the window. "Control Effort" shows the output of the controller. So, you can decide whether your control can produce such a signal or not.

In order to understand the meaning of plots drawn by PID Tuner, we draw the following simulation diagram in Simulink®. We use a "PID Controller" block shown in Fig. 4.20. The "PID Controller" block setting is shown in Fig. 4.21. If we run the simulation we obtain the waveform shown in Figs. 4.22 and 4.23. Compare these figures with Fig. 4.18.

The last step is to test the designed controller. We set up the following simulation diagram. Output voltage and PID controller output is shown in Figs. 4.25 and 4.26, respectively. As we expect, the duty cycle is about 40% for 20 V output.

We use the simulation diagram shown in Fig. 4.27 to test the designed controller. We assume that input voltage is changed from 50–60 V at $t = 25$ ms. This is done with the aid of the left "Step" block. The setting of the block is shown in Fig. 4.28.

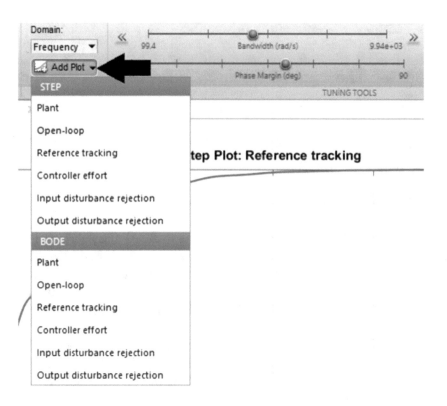

Figure 4.17: "Add Plot" button.

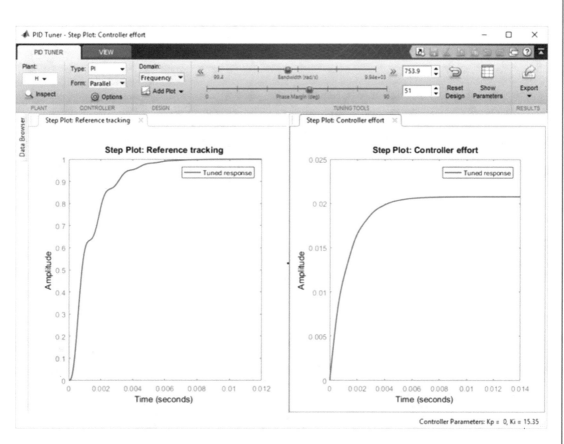

Figure 4.18: Adding plot to the PID Tuner window.

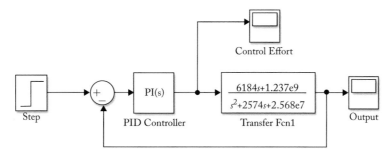

Figure 4.19: Simulink®diagram of the closed-loop system.

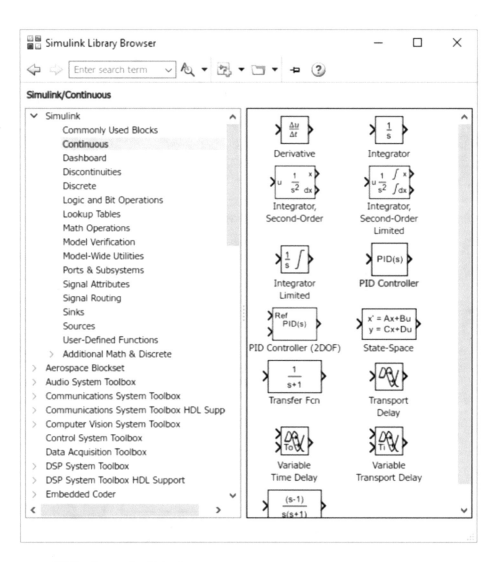

Figure 4.20: "PID Controller" block.

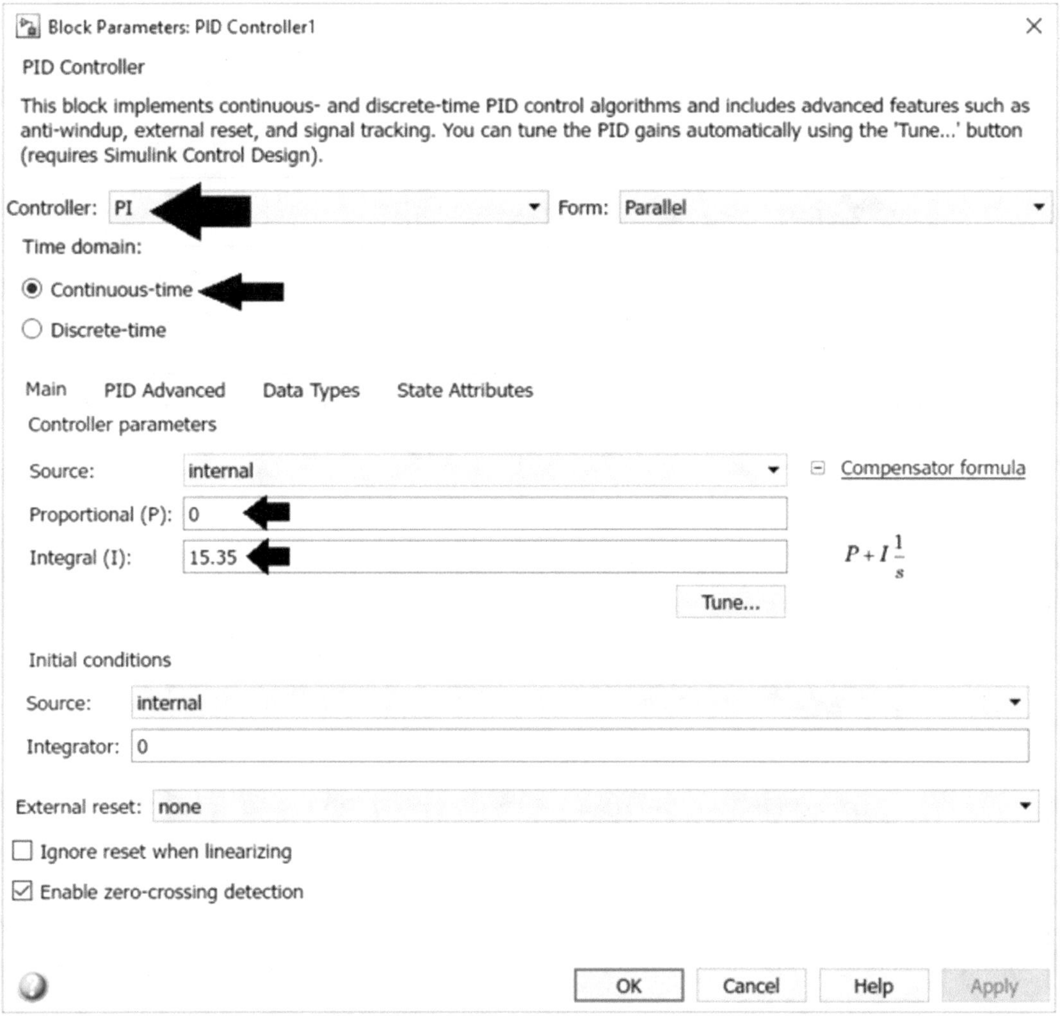

Figure 4.21: "PID Controller" block (Fig. 4.19) settings.

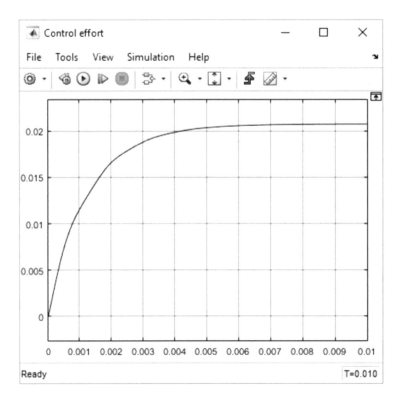

Figure 4.22: Controller output signal (see Fig. 4.19).

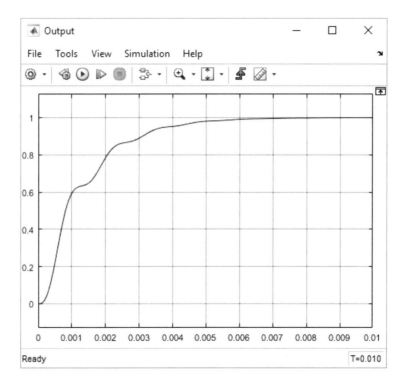

Figure 4.23: Output signal (see Fig. 4.19).

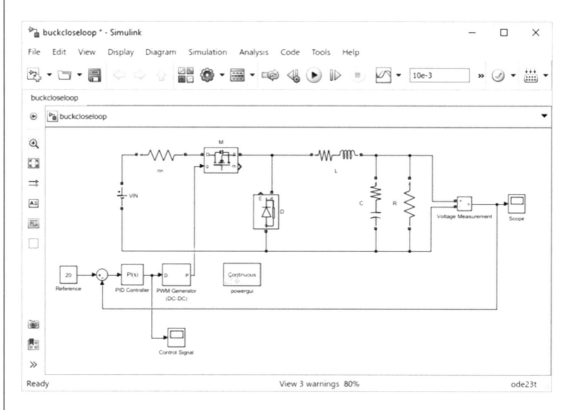

Figure 4.24: Simulink®diagram of a closed-loop Buck converter.

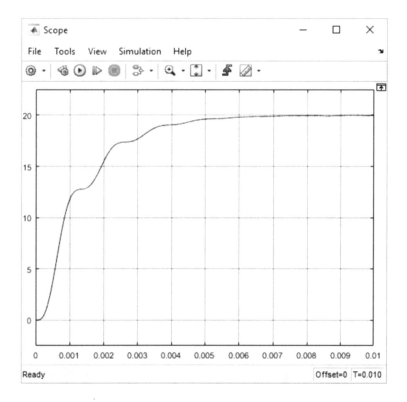

Figure 4.25: Output voltage of a closed-loop Buck converter.

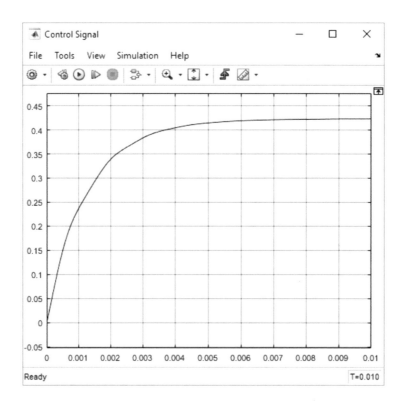

Figure 4.26: Output of a controller (see Fig. 4.24).

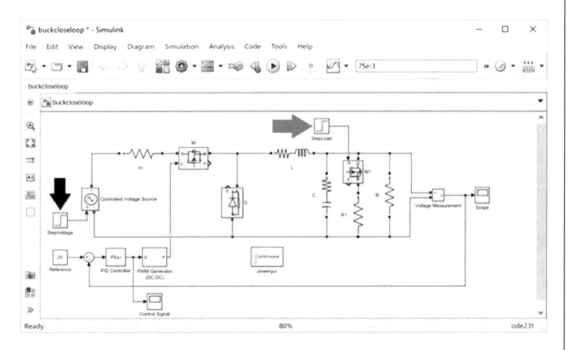

Figure 4.27: Simulation diagram for studying the effect of disturbances.

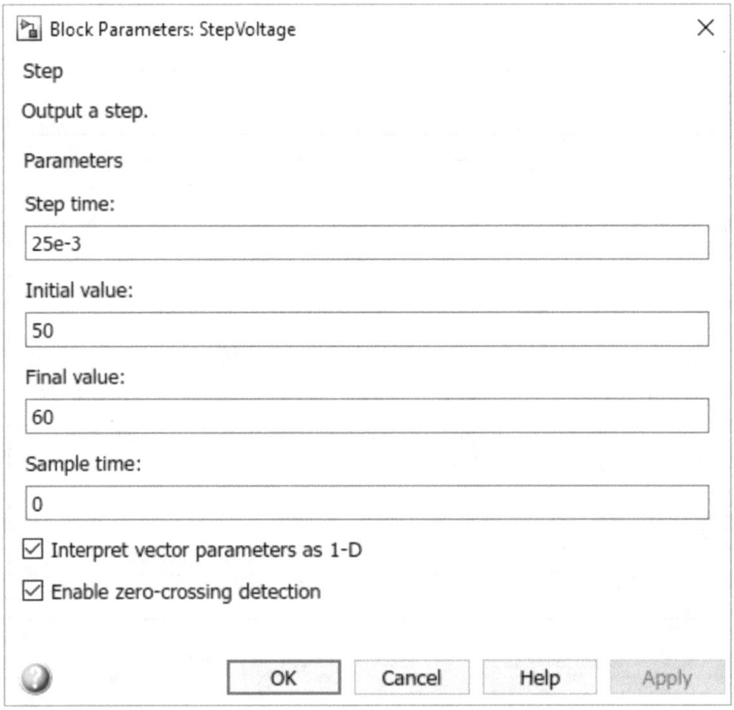

Figure 4.28: Step block settings (shown with black arrow in Fig. 4.27).

Load changes from $R = 5\ \Omega$ to $\frac{5\times1}{5+1} = 0.83\ \Omega$ at $t = 50$ ms (Resistor $R1$ value is 1 Ω) with the aid of "Step" block and a MOSFET. The setting of "Step" block is shown in Fig. 4.29. When output of "Step" block goes to one, the MOSFET closes and cause the resistor $R1$ to be in parallel with resistor R. This causes the value of load to change suddenly. Simulation results are shown in Fig. 4.30. As you see the controller keeps the output constant despite disturbances.

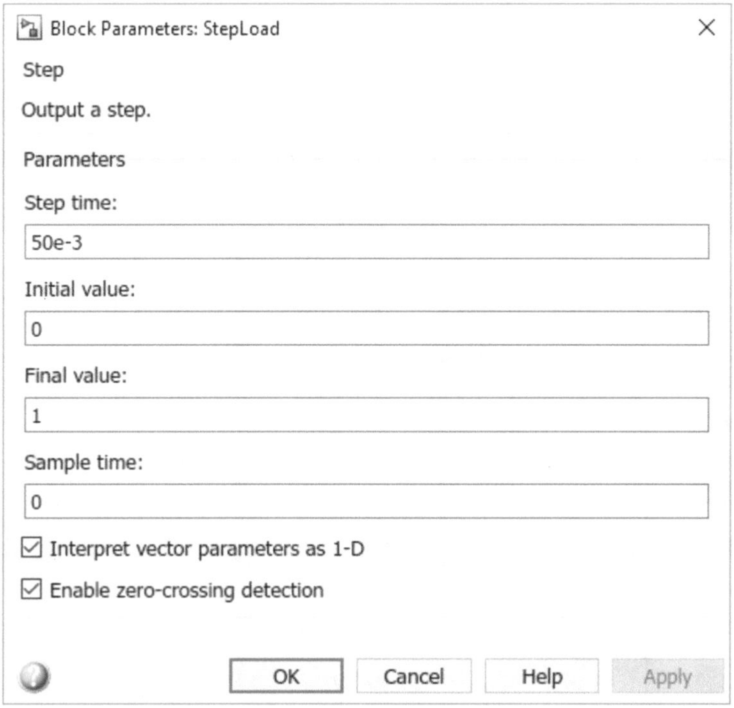

Figure 4.29: Step block settings (shown with grey arrow in Fig. 4.27).

We can even study the effect of change in reference. The simulation diagram is changed, as shown in Fig. 4.31. Just a step block has been added to the previous diagram. The reference signal is taken from a "Step" block with setting shown in Fig. 4.32. Reference changes from 20–30 V at $t = 75$ ms. Simulation results are shown in Fig. 4.33. As you can see, the controller tracks the reference signal. Figure 4.34 shows the output of PID controller. As you see, controller changes the duty ratio so the output tracks the reference despite disturbances.

You can use the "sisotool" to design the controller instead of "pidTuner." "sisotool" provides more options. You can use the commands shown in Fig. 4.35 to enter the sisotool environment. The sisotool environment is shown in Fig. 4.36.

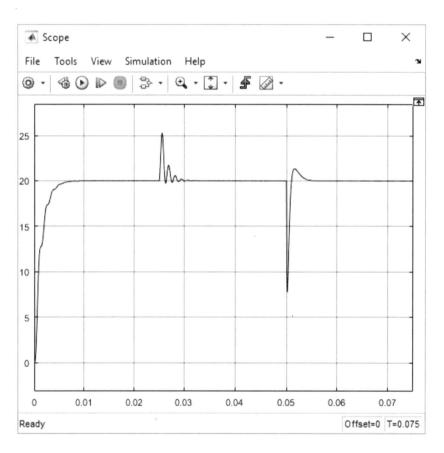

Figure 4.30: Simulation result.

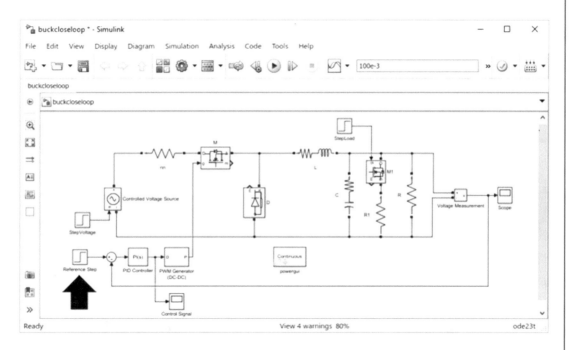

Figure 4.31: Simulation diagram to study the effect of change in the reference signal.

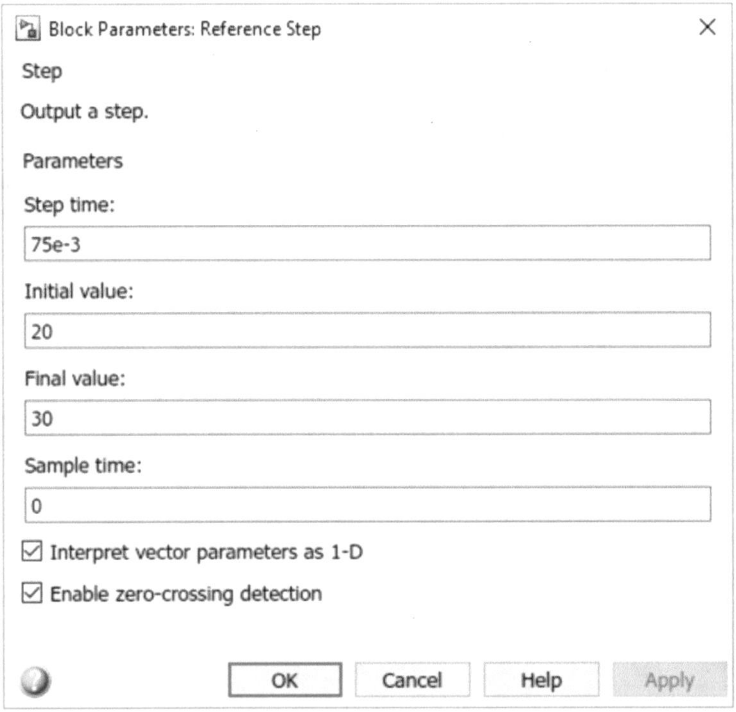

Figure 4.32: Step block settings (shown with black arrow in Fig. 4.31). Other blocks are the same as the previous simulation.

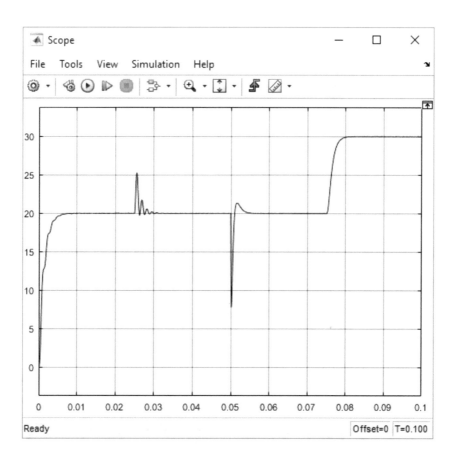

Figure 4.33: Simulation result.

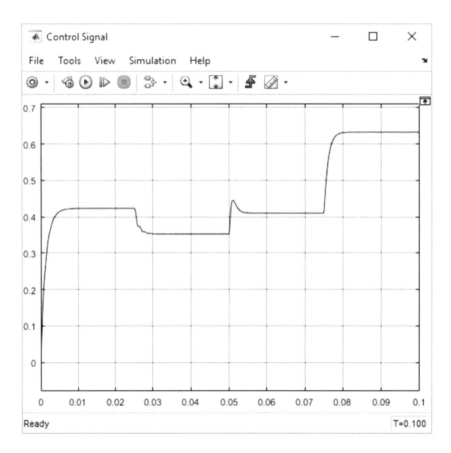

Figure 4.34: Controller output.

```
Command Window
   >> H=tf([6184 1.237e9], [1 2574 2.568e7]);
   >> sisotool(H)
fx >>
```

Figure 4.35: Running the sisotool.

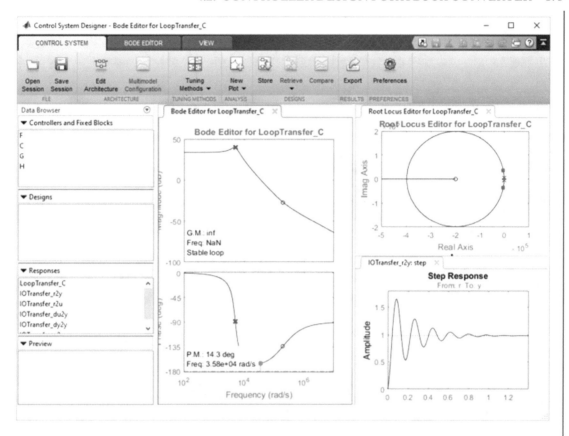

Figure 4.36: "Control System Designer" window is shown after running the "sisotool" command.

You can select the desired structure using the "Edit Architecture" button as shown in Fig. 4.37. Normally, the first option is used. Using the icon ⬇ you can define the transfer function of blocks in the loop.

You can use the "Tuning Methods" button (Fig. 4.38) and "Automated Tuning" section to automatically tune the controller (Block C in Fig. 4.37). If you click the "PID Tuning," a window like that shown in Fig. 4.39 appears. You can use the slider to tune the PID controller as done in previous sections.

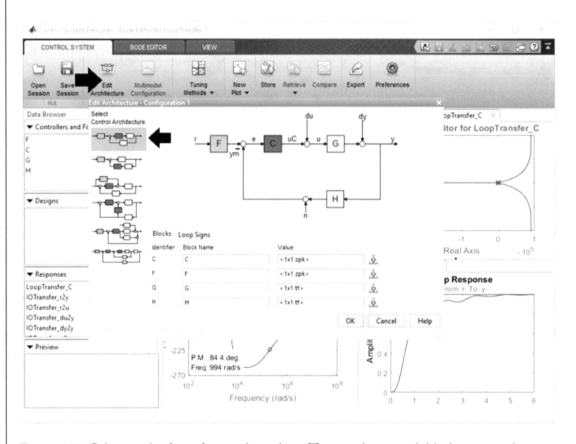

Figure 4.37: Selecting the desired control topology. The one shown with black arrow is the most common topology.

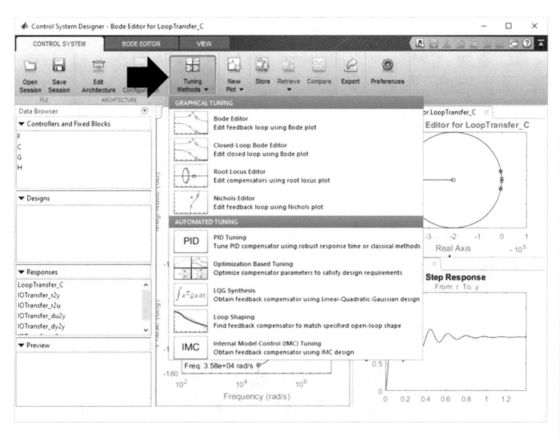

Figure 4.38: "Tuning Methods" button.

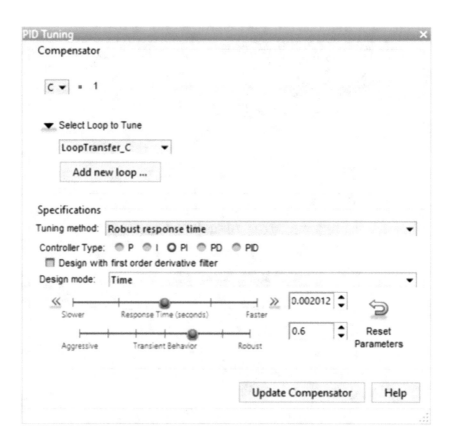

Figure 4.39: **PID Tuning** window.

4.3   LOOPSHAPING

You can use design the controller by using the loop shaping method. Assume a control structure like that shown in Fig. 4.40. $C(s)$ and $P(s)$ show the controller and plant transfer function, respectively.

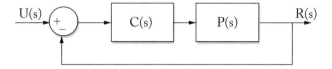

Figure 4.40: Closed-loop control.

The closed-loop transfer function is:

$$H_{CL}(s) = \frac{R(s)}{U(s)} = \frac{C(s)P(s)}{1 + C(s)P(s)}.$$

The product of $C(s) \times P(s)$ is called "loop transfer function." Assume that $C(s)P(s) = \frac{K}{s}$. In this case, the closed-loop transfer function is

$$H_{CL}(s) = \frac{R(s)}{U(s)} = \frac{\frac{K}{s}}{1 + \frac{K}{s}} = \frac{K}{s + K}.$$

If we apply a step function to such a system, it tracks the step function with zero steady state error within $\frac{5}{K}$ s. For example, if $C(s)P(s) = \frac{1000}{s}$, it takes about $\frac{5}{1000} = 5$ ms, to track the step command. The response is shown in Fig. 4.41. Note that the response has no overshoot or oscillatory nature.

As another example, assume that $C(s)P(s) = \frac{\omega_n^2}{s(s+2\zeta\omega_n)}$. In this case, the closed-loop transfer function is

$$H_{CL}(s) = \frac{R(s)}{U(s)} = \frac{\frac{\omega_n^2}{s(s+2\zeta\omega_n)}}{1 + \frac{\omega_n^2}{s(s+2\zeta\omega_n)}} = \frac{\omega_n^2}{s^2 + 2\zeta\omega_n s + \omega_n^2}.$$

Based on the values of ζ and ω_n, the response takes different forms. Figure 4.42 shows the step response for two different values of ζ. As you see, an increase in ζ decreases the oscillatory nature.

Loop-shaping takes the desired loop transfer function (i.e., $H_{desired}(s)$) and design the controller block (block C in Fig. 4.40) such that the error between $H_{desired}(s) - C(s)P(s)$ is minimized.

One may ask, $C(s) = \frac{H_{desired}(s)}{P(s)}$ makes the error zero since

$$H_{desired}(s) - C(s)P(s) = H_{desired}(s) - \frac{H_{desired}(s)}{P(s)}P(s) = 0.$$

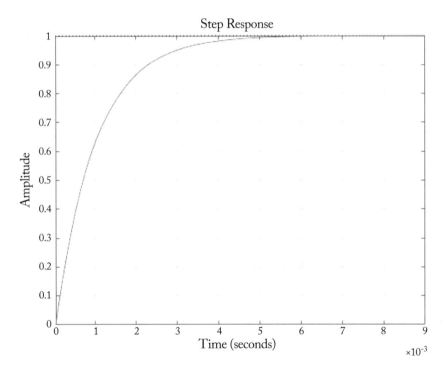

Figure 4.41: Step response of $\frac{1000}{s+1000}$.

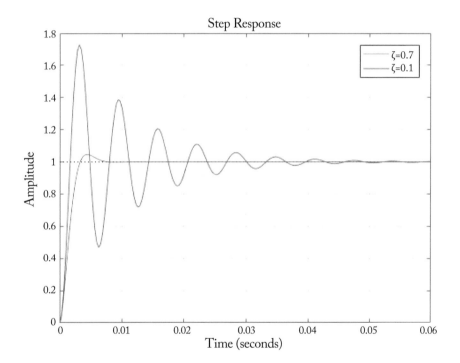

Figure 4.42: Effect of damping ratio (ζ) on the step response of a second-order system.

Although the approach seems correct, it is not applicable except for simple cases. Studying an example is quite usefull. For example, assume that $P(s) = \frac{4}{(s+7)(s+8)}$ and $H_{desired}(s) = \frac{10}{s}$. So, $C(s) = \frac{H_{desired}(s)}{P(s)} = \frac{10(s+7)(s+8)}{4s}$. The obtained controller is not proper so it is not realizable.

The loop-shaping button in the "Automated Tuning" section of Fig. 4.38 uses optimization techniques to obtain the proper controller $C(s)$. The user translates the requirement into the $H_{desired}(s)$, for example if one need a response free of overshoot with settling time less than 5 ms he/she may use $H_{desired}(s) = C(s)P(s) = \frac{1000}{s+1000}$ is a good option. If the overshoot of less than 5% is acceptable one may use $H_{desired}(s) = C(s)P(s) = \frac{1000^2}{s(s+2\times0.7\times1000\times s)}$. Normally, $H_{desired}(s)$ is selected among the first- and second-order transfer functions to avoid increase in the order of designed controller.

When you click the "Loop Shaping" button in Fig. 4.38 the window shown in Fig. 4.43 appears. You enter the selected $H_{desired}(s)$ into the "Target open-loop shape (LTI)" box using the command "tf". "Frequency range for loop shaping [wmin,wmax]:" box takes the frequency portion where the optimization must be done. Normally, we need the overlap in the low frequency range.

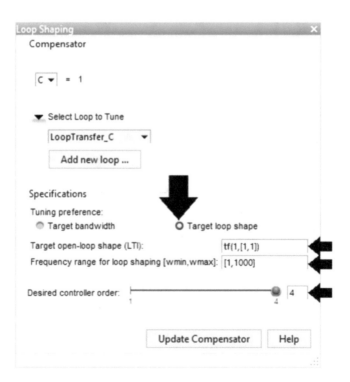

Figure 4.43: "Loop Shaping" window.

"Desired controller order" slider bar takes the controller order. You start with low order controllers (i.e., a first order controller). The design process is started by pressing the "Update Compensator" button. When the order of the controller is too low, the software shows a warning and ask you to increase the order. In this case, you increase the order by one. If the warning comes out again, you increase it again.

You can transfer the designed controller to MATLAB's workspace by clicking on the "Export" button (Fig. 4.44) in the sisotool's main window. After clicking the button, the "Export Model" window pops up, as shown in Fig. 4.45. You select the desired block and click on "Export" button to export the block to workspace. For instance, if you want to export the designed controller you select the block named "C" and press the "Export."

You can use other automatic design methods like LQG or IMC to design controllers (see Fig. 4.38) as well.

4.4   OLDER VERSIONS OF SISOTOOL

If you use older versions of MATLAB®(i.e., R2015), you see a user interface like that shown in Fig. 4.46 when you enter the sisotool environment. In this case, you must click on "Automated Tuning" tab to access different automatic controller design methods.

4.5 MANUAL CONTROLLER DESIGN

You can add your desired pole/zero to the controller. Adding the pole/zero can be done by right clicking one of the diagrams shown in the sisotool environment (Fig. 4.48). You can add the desired pole/zero by selecting the "Add Pole/Zero," as shown in Fig. 4.49. As shown in Fig. 4.49, you can add integrator, lead, lag, etc. easily by clicking the corresponding choice.

You can refer to the references at the end of this chapter to learn manual controller design.

Figure 4.44: "Export" button.

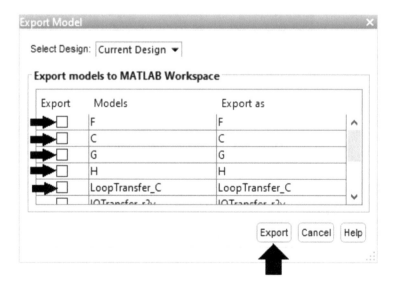

Figure 4.45: Exporting the desired transfer function is possible by selecting it and pressing the "Export" button.

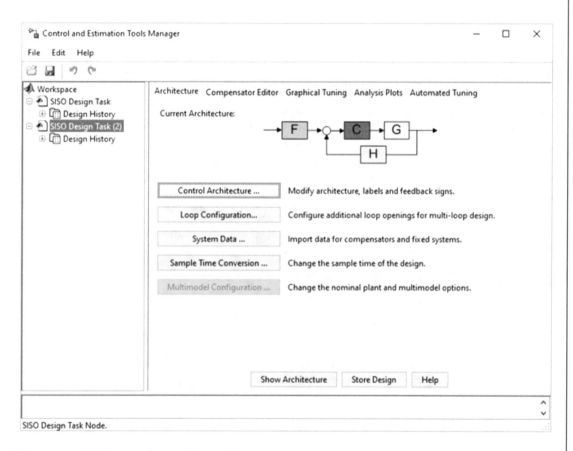

Figure 4.46: Older versions of sisotool.

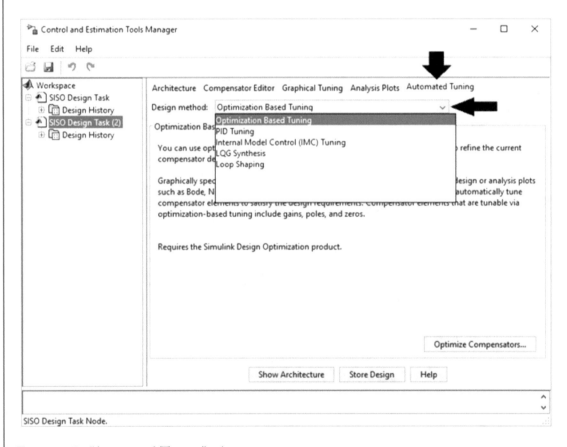

Figure 4.47: "Automated Tuning" tab.

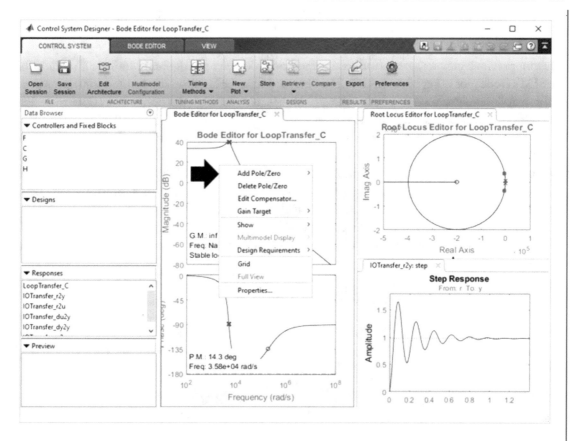

Figure 4.48: Right clicking on the plot opens the menu.

Figure 4.49: "Add Pole/Zero" menu.

4.6 CONTROLLER DESIGN BASED ON SYSTEM IDENTIFICATION

In the previous chapter, we introduced a method to extract an algebraic transfer function for PSIM's Bode plot using MATLAB®. The obtained transfer function can be used for the purpose of controller design. So, one may use the techniques introduced in the previous sections to design a controller after he/she obtains the model.

In this section we introduce another method to design controllers. A Boost converter working in DCM is chosen as example but the method can be used for CCM converters as well. We use the System Identification Toolbox™of MATLAB®. System identification tries to extract a model for a system based on its input-output signals. For more information about system identification, refer to the references at the end of chapter.

We want to design a controller for the Boost converter shown in Fig. 4.50. rC and rL show capacitor and inductor Equivalent Series Resistance (ESR), respectively.

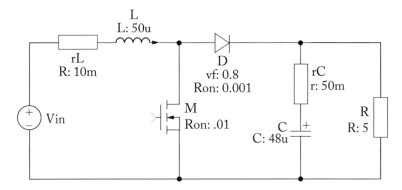

Figure 4.50: Boost converter.

The simulation diagram shown in Fig. 4.51 is used to simulate the circuit. The switching frequency is 25 KHz. Assume that output must be about 30 V. We change the duty ratio of MOSFET gate signal to obtain the 30 V output. After some trial and error, duty ratio (D) of $D = 0.46$ provide the output of 30 V.

The inductor current steady state current is shown in Fig. 4.52 for $D = 0.46$. As you see, the converter works in Discontinuous Current Mode (DCM). The output voltage is shown in Fig. 4.53. If we zoom in on the output voltage, we see the output ripple, as shown in Fig. 4.54.

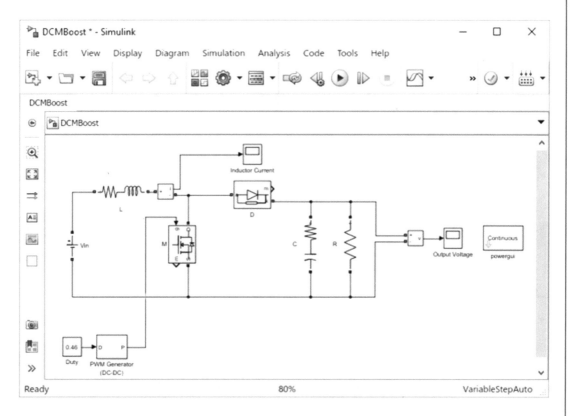

Figure 4.51: Simulation diagram of the Boost converter in the Simulink®environment.

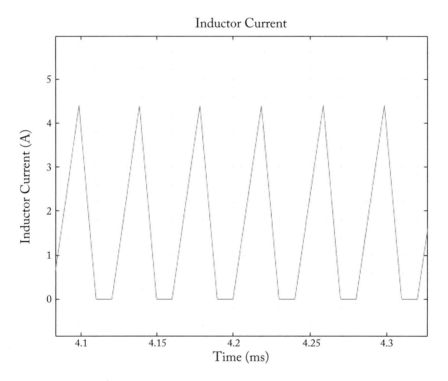

Figure 4.52: Inductor current (see Fig. 4.51).

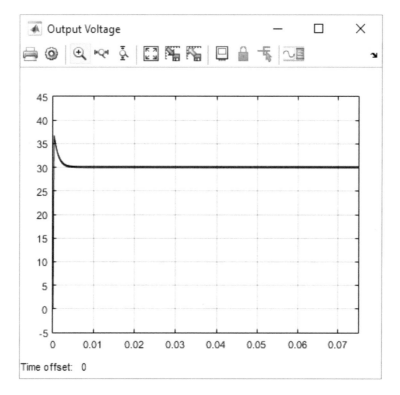

Figure 4.53: Output voltage (see Fig. 4.51).

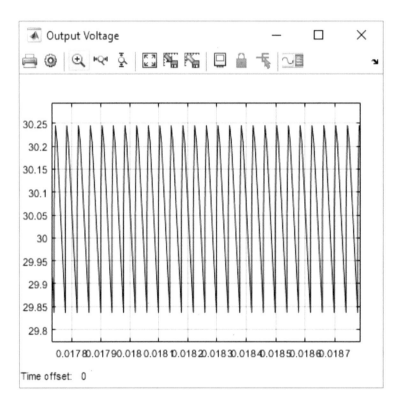

Figure 4.54: A closer loop at output voltage.

We want to obtain the output voltage's average value. To do this we add a "Mean" block to the simulation file, as shown in Fig. 4.55. "Mean" block is obtainable from the "Measurements" section, as shown in Fig. 4.56. To obtain the average output waveform, the "Mean" block's "Fundamental frequency (Hz)" box must contain the same value as the switching frequency. The switching frequency is 25 KHz (see Fig. 4.57) so we fill the "Fundamental frequency (Hz)" box with 25,000, as shown in Fig. 4.58.

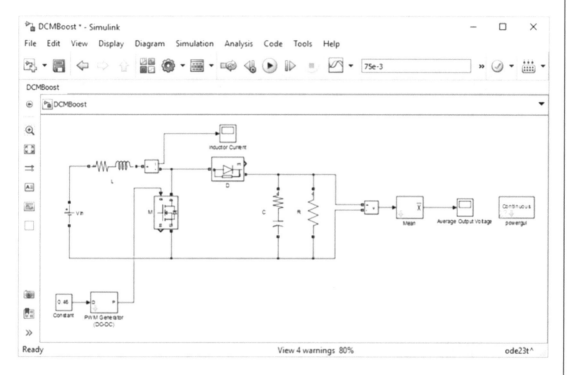

Figure 4.55: Adding the "Mean" block to the simulation diagram.

Output of "Mean" block is shown in Fig. 4.59. If we zoom in the waveform we see no ripple. Compare Fig. 4.60 with 4.54.

We need to transfer the obtained average waveform to the MATLAB®workspace for further processing. This can be done with the aid of "To Workspace" block. We change the simulation file to that shown in Fig. 4.61. "To Workspace" block can be found under the "Sinks" section, as shown in Fig. 4.62.

We do the "To Workspace" block settings, as shown in Fig. 4.63. This causes the block to send a value to MATLAB®workspace every 10^{-5} s.

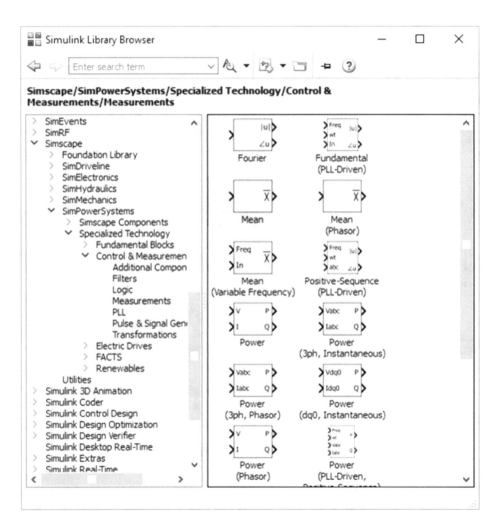

Figure 4.56: "Mean" block.

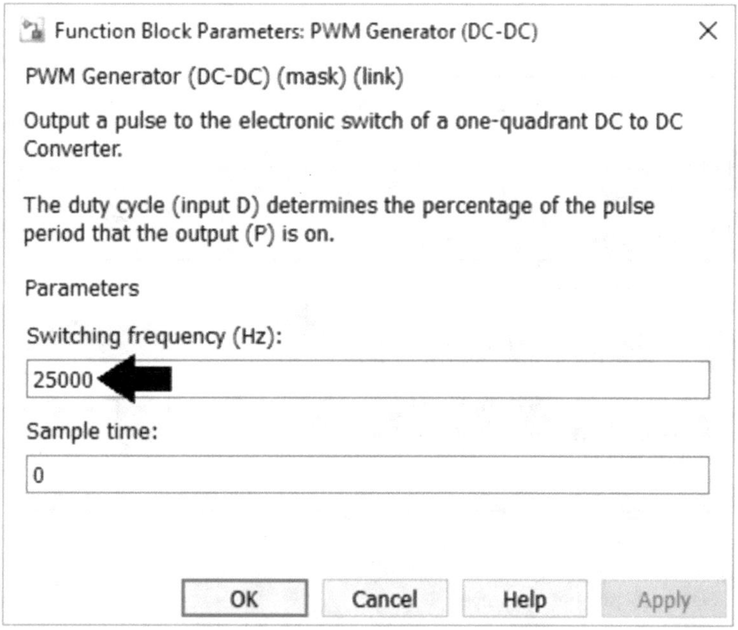

Figure 4.57: "PWM Generator (DC-DC)" block settings.

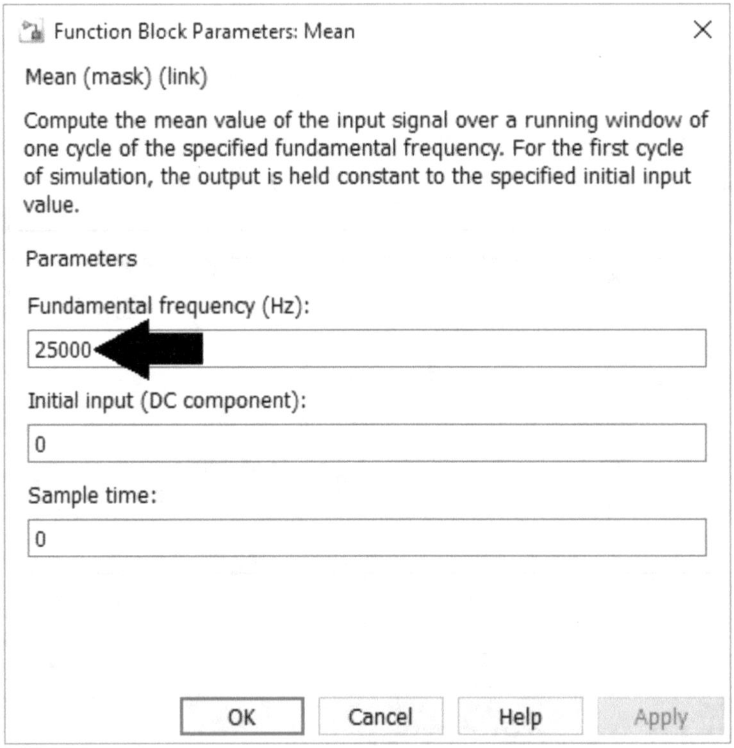

Figure 4.58: "Mean" block settings.

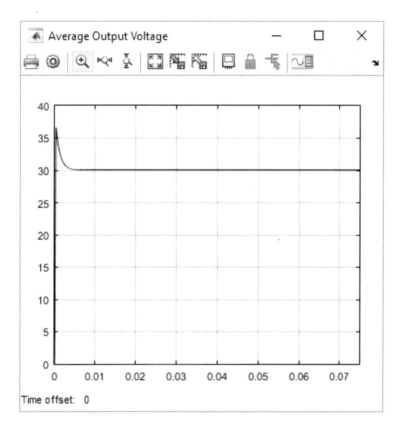

Figure 4.59: Scope "Average Output Voltage" waveform (see Fig. 4.55).

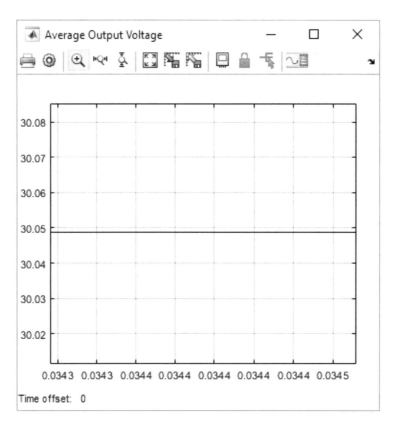

Figure 4.60: A closer look at scope "Average Output Voltage" waveform.

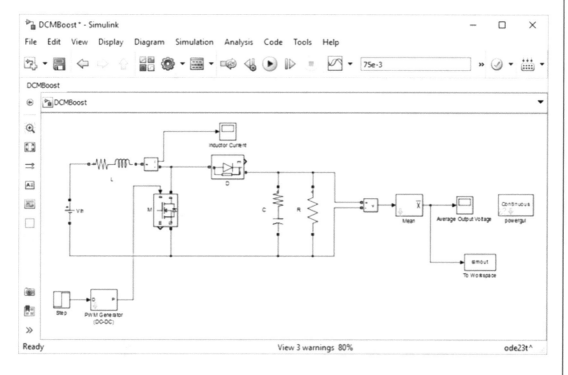

Figure 4.61: Adding the "To Workspace" block.

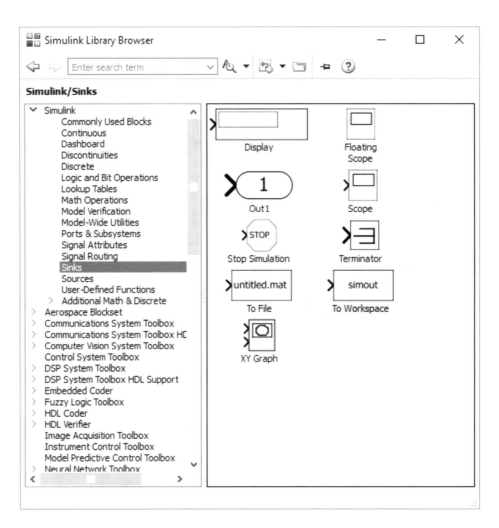

Figure 4.62: The "To Workspace" block.

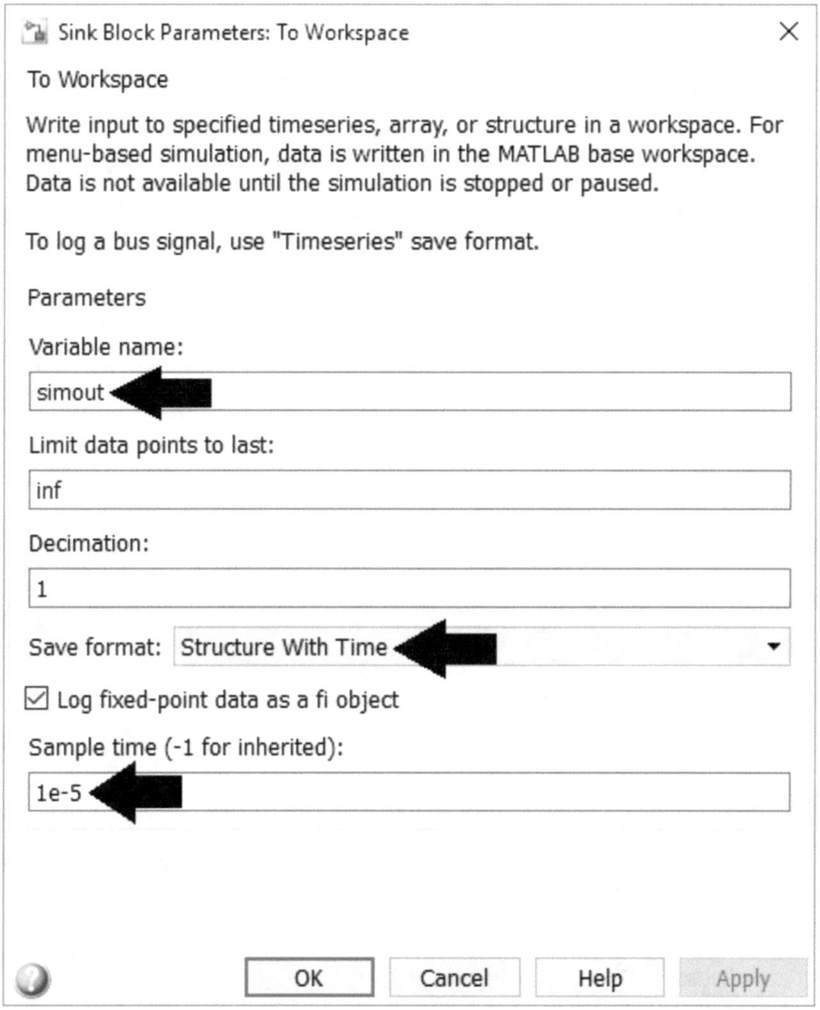

Figure 4.63: The "To Workspace" block settings.

As shown in Fig. 4.61, "PWM Generator (DC-DC)" input is connected to a step block. The step block setting is shown in Fig. 4.64. The step blocks set the duty ratio of MOSFET gate signal. For $0 < t < 50$ m, the step block produces 0.46 and for $t > 50$ ms it produces 0.47. This small change acts as perturbation and helps us to extract the control-to-output transfer function. In fact, we stimulate the control input (duty ratio) with a step function of amplitude $0.47 - 0.46 = 0.01$.

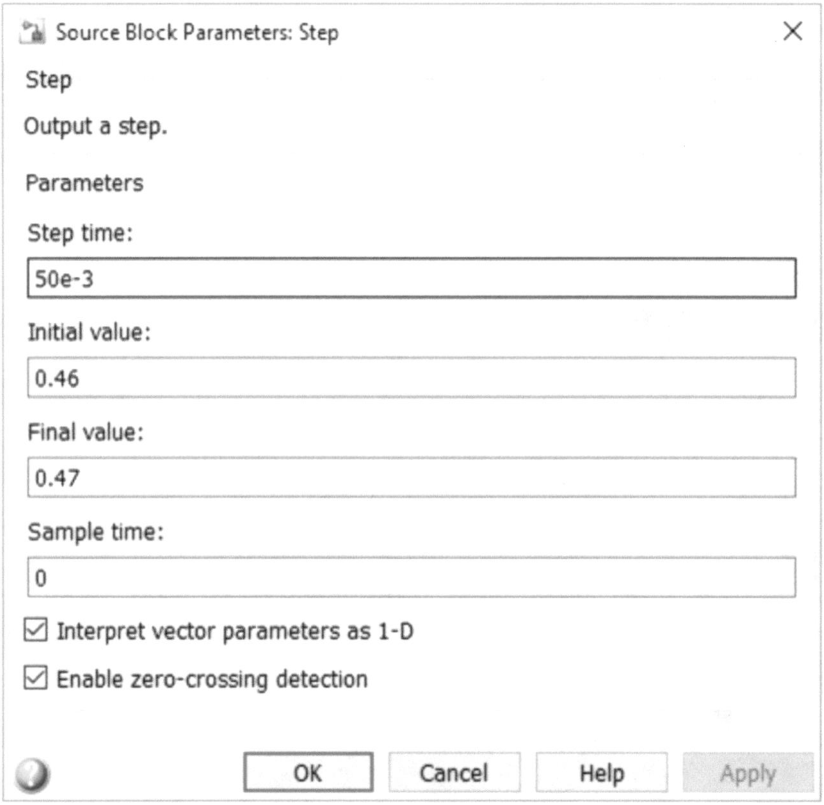

Figure 4.64: "Step" block settings.

If we run the simulation we can see the change in output average value caused by change in duty ratio (Fig. 4.65). After the simulation is done, you can see some new variables in the workspace (Fig. 4.66). These variables are produced by the "To Workspace" block. We extract the information in the variable named "simout" into two new variables named t and *Vout*.

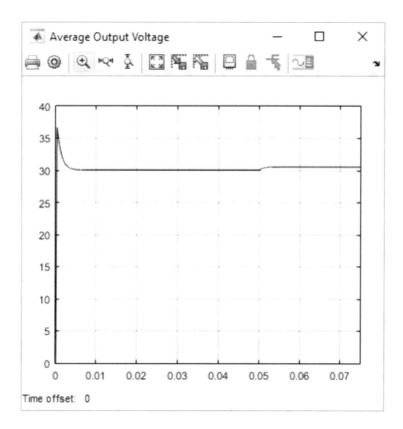

Figure 4.65: Simulation result.

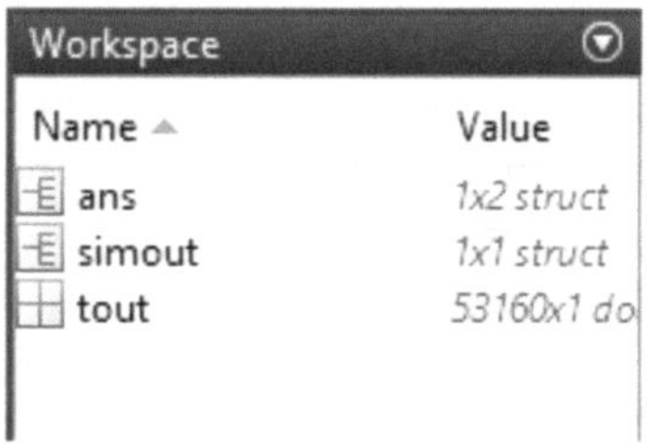

Figure 4.66: Variables named "simout" and "tout" are added to the workspace.

t contains "time" information (i.e., horizontal axis of Fig. 4.65) and *Vout* contains average output values (i.e., vertical axis of Fig. 4.65). If we plot *Vout* as shown in Fig. 4.68, we obtain Fig. 4.69 which is the same as Fig. 4.65. As you can see, the output average value increased for $t \geq 50$ ms. A closer look is shown in Fig. 4.70.

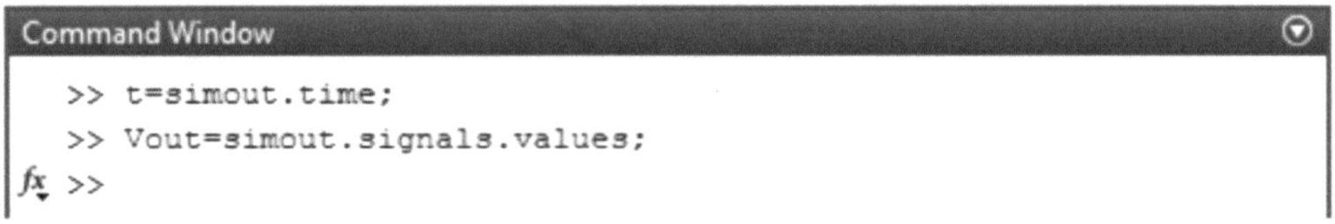

```
Command Window                                              ⊙
    >> t=simout.time;
    >> Vout=simout.signals.values;
fx >>
```

Figure 4.67: Extracting the data from variable named "simout" into a variable named "Vout."

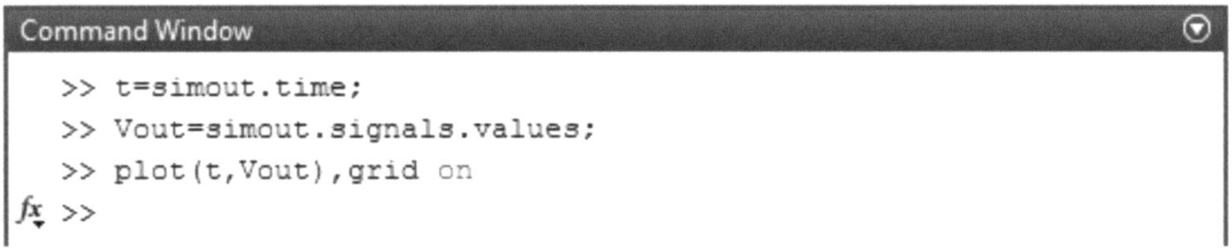

```
Command Window                                              ⊙
    >> t=simout.time;
    >> Vout=simout.signals.values;
    >> plot(t,Vout),grid on
fx >>
```

Figure 4.68: Plotting the data.

We need to obtain the $t \geq 50$ ms portion of the graph. Since duty ratio starts to change at $t = 50$ ms. The absolute value of averaged output is *not* important for transfer function extraction, the change in averaged out is important. We use the code shown in Fig. 4.71 to extract the $t \geq 50$ ms portion of the graph. The extracted section is plotted in Fig. 4.72.

We want to find a suitable transfer function for the control-to-output based on the obtained signals from simulation. MATLAB's System Identification Toolbox™can extract the transfer function for us. We type "ident" in order to call the Toolbox™. This is shown in Fig. 4.73. After pressing the Enter key, a window like that shown in Fig. 4.74.

We select "Time domain data" from the drop-down list, as shown in Fig. 4.75. After clicking the "Time domain data," the "Import Data" window will be opened. We fill the window as shown in Fig. 4.76. The "Input" box must contain the signal which used to stimulate the system. We stimulated the system with step of amplitude $0.47 - 0.46 = 0.01$. So, we used ".01*ones(size(x))" which produce a vector which all the elements are 0.01. "Output" is the signal produced by the system after stimulation is applied. We transfer the system response to a variable named y, so we fill the "Output" box with letter y. "Sample time" box must be filled with the same value which is used for producing the data before. The "To Workspace" block sample the signal every 10^{-5} s (Fig. 4.63) so the "Import Data" window "Sample time" box is filled with 1e-5 $= 10^{-5}$. After filling the required data, the "Import" and "Close" buttons are clicked.

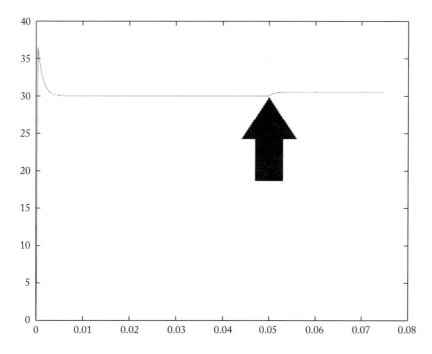

Figure 4.69: Output starts to change at $t = 50$ ms.

30.8
30.7
30.6
30.5
30.4
30.3
30.2
30.1
30
29.9
29.8
0.05
0.055
0.06
0.065
0.07

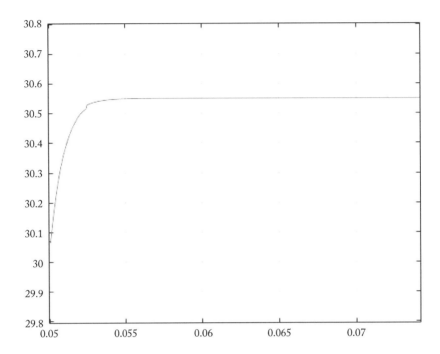

Figure 4.70: A closer look at the $t > 50$ ms portion of the graph in Fig. 4.69.

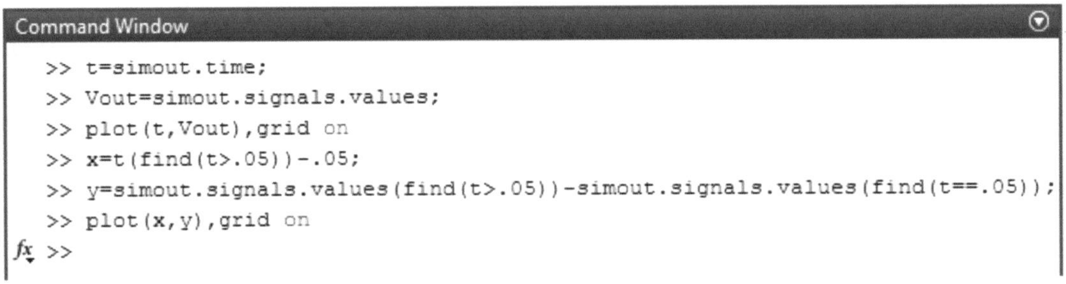

Command Window
>> t=simout.time;
>> Vout=simout.signals.values;
>> plot(t,Vout),grid on
>> x=t(find(t>.05))-.05;
>> y=simout.signals.values(find(t>.05))-simout.signals.values(find(t==.05));
>> plot(x,y),grid on
fx >>

Figure 4.71: Removing the large signal portion.

0.5
0.4
0.3
0.2
0.1
0
-0.1
0
0.005
0.01
0.015
0.02
0.025
0.03

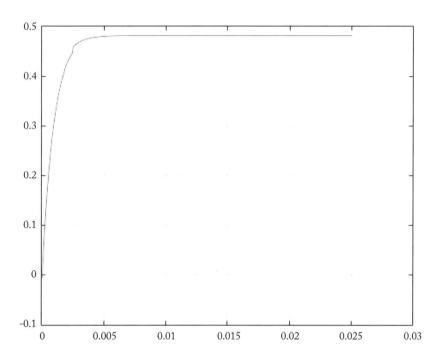

Figure 4.72: Change in the output voltage caused by change in duty ratio.

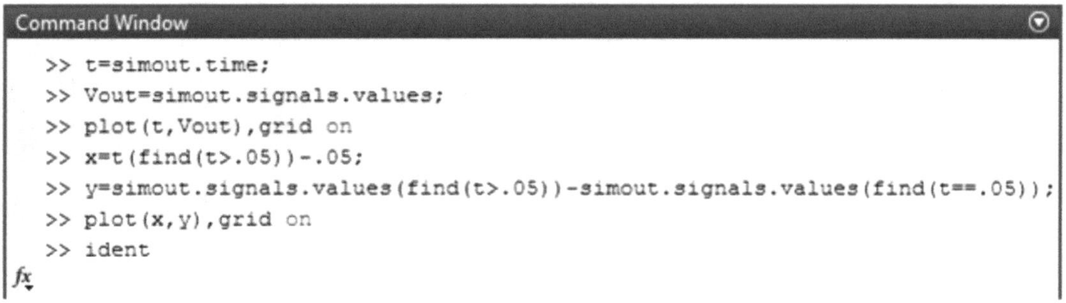

Command Window
>> t=simout.time;
>> Vout=simout.signals.values;
>> plot(t,Vout),grid on
>> x=t(find(t>.05))-.05;
>> y=simout.signals.values(find(t>.05))-simout.signals.values(find(t==.05));
>> plot(x,y),grid on
>> ident
fx

Figure 4.73: Running the "System Identification" Toolbox™. The command "ident" runs the Toolbox™.

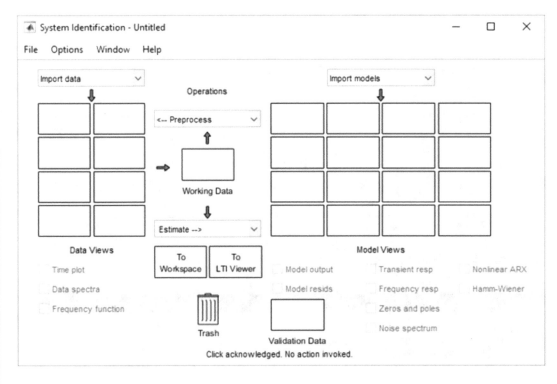

Figure 4.74: "System Identification" window.

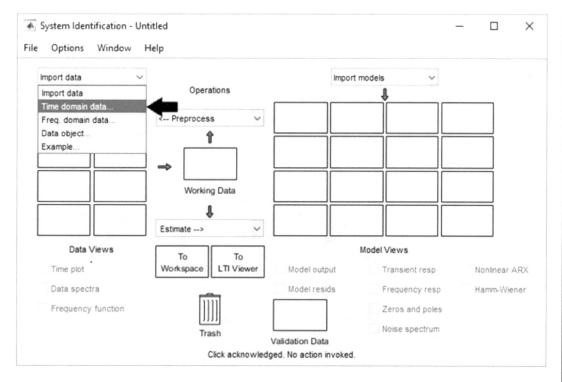

Figure 4.75: Importing the data into the Toolbox™.

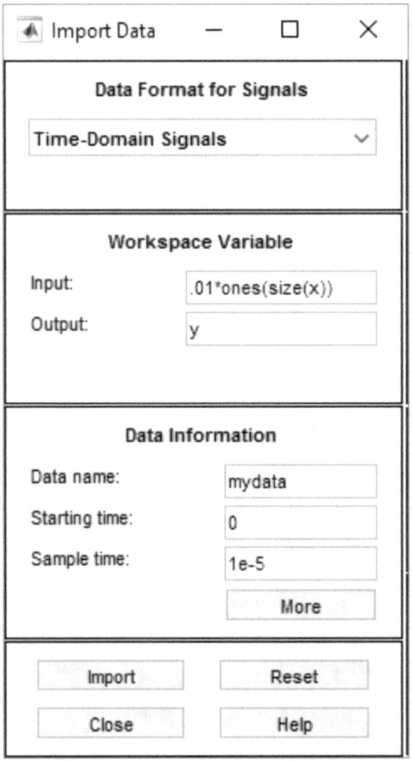

Figure 4.76: "Import Data" window.

You can see the time plot of imported data by clicking the "Time plot" check box (Fig. 4.77). The plot of imported data is shown in Fig. 4.78. This plot is the same as Fig. 4.72.

The transfer function can be extracted by clicking on the "Transfer Function Models," as shown in Fig. 4.79. After clicking the "Transfer Function Models," the window shown in Fig. 4.80 appears. Since we want a continuous model, "Continuous-time" is selected. The boost converter contains two dynamic elements, namely inductor and capacitor. So, we expect the transfer function of order two (i.e., two poles).

When the "Estimate" button is clicked, MATLAB®extracts the best transfer function of order two to the given data. After some calculation, results are shown in a window like that shown in Fig. 4.81. You must notice to the "Fit to estimation data" part of the window. It shows the goodness of obtained transfer function. In this case, it reaches 98.37% so it is reliable. If the "Fit to estimation data" is a low number you must redo the previous steps (see Figs. 4.79 and 4.80). "Number of poles" and especially "Number of zeros" are changed to obtain a better result. Generally, the "Number of poles" is equal to the number of inductors plus capacitors in the converter.

If you click the "Model output" (Fig. 4.82) you can see the calculated model output and output signal on the same graph. This makes comparison easy. As shown in Fig. 4.83, model output and output signal are overlapped so the calculated model is a good representative of the system. You can see the calculated models zeros and poles by clicking the "Zeros and poles" checkbox. The calculated model's zeros and poles are shown in Fig. 4.84.

You can export the calculated transfer function to the MATLAB®workspace by drag and drop. To do this, we drag the "tf1" shown in Fig. 4.85 and drop it on the "To Workspace" icon. After drag and drop, a variable named "tf1" is available in the workspace, as shown in Fig. 4.86. We can use this model to design a controller. Assume that you want to design a PI controller for this Boost converter. You enter the command "pidTuner(tf1)," as shown in Fig. 4.87. The window shown in Fig. 4.88 is opened and you change the slider gains to obtain the desired response.

After obtaining the desired response, we set up a simulation like that shown in Fig. 4.89. In this case, we used a controller with the parameters shown in Fig. 4.90. As shown in Fig. 4.89, the reference of control system is taken from a "Step" block. The "Step" block settings are shown in Fig. 4.91. The control systems reference is 30 V for $0 \leq t \leq 50$ ms and is 40 V for $t \geq 50$ ms. Simulation results are shown in Fig. 4.92. The boost converter tracks the reference, as shown in Fig. 4.92.

You can use different scenarios to test the designed controller. For example, in Fig. 4.93 effect of change in input voltage has been studied. The input voltage is controlled by a step source. Step block setting is shown in Fig. 4.94. Reference of the control system is constant of 30 V. The simulation result is shown in Fig. 4.95. As you see, the controller keeps the voltage constant despite change in input voltage.

Figure 4.77: "Time plot" check box.

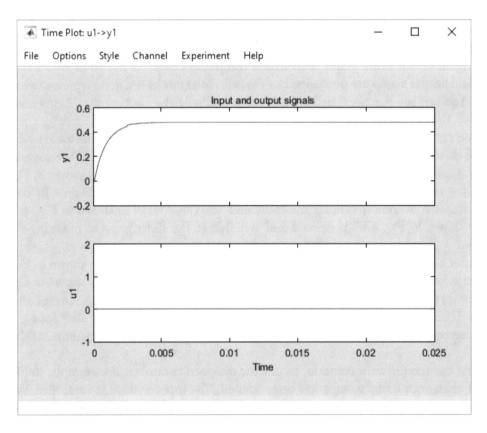

Figure 4.78: Graph of input and output signals.

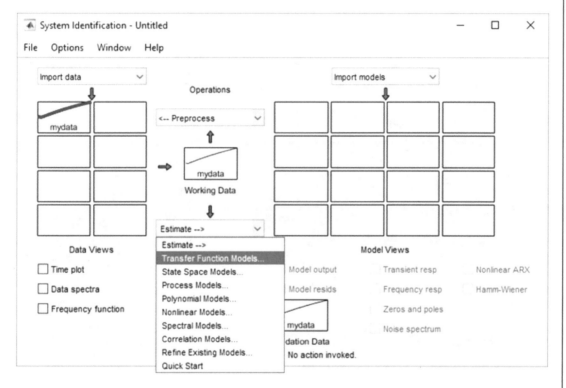

Figure 4.79: Estimating a transfer function for imported data.

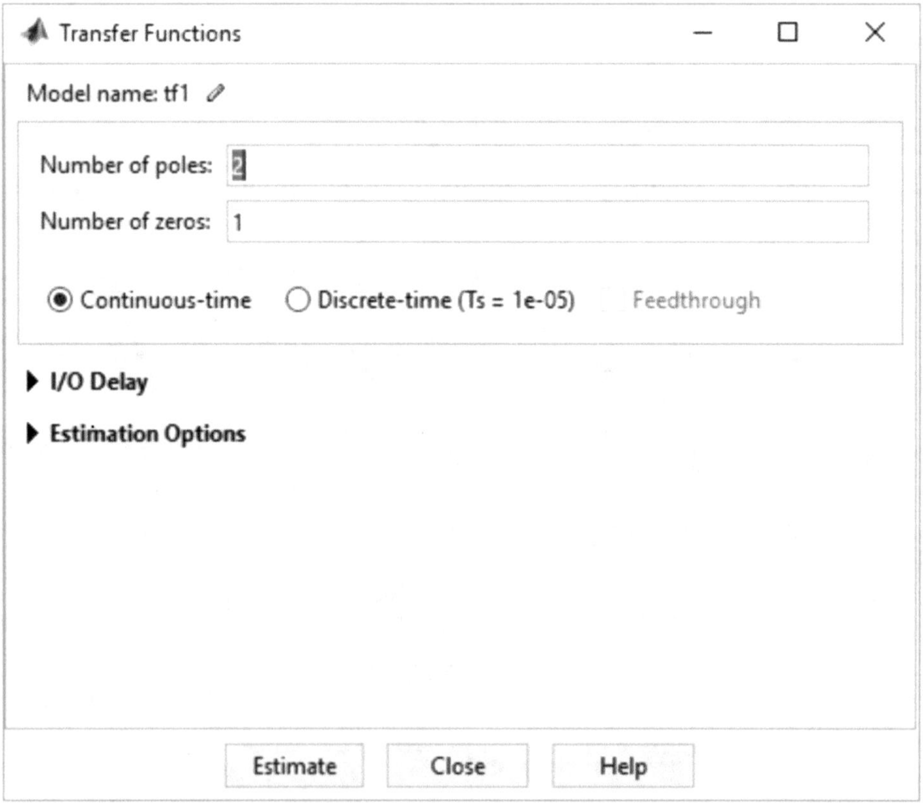

Figure 4.80: "Transfer Functions" window.

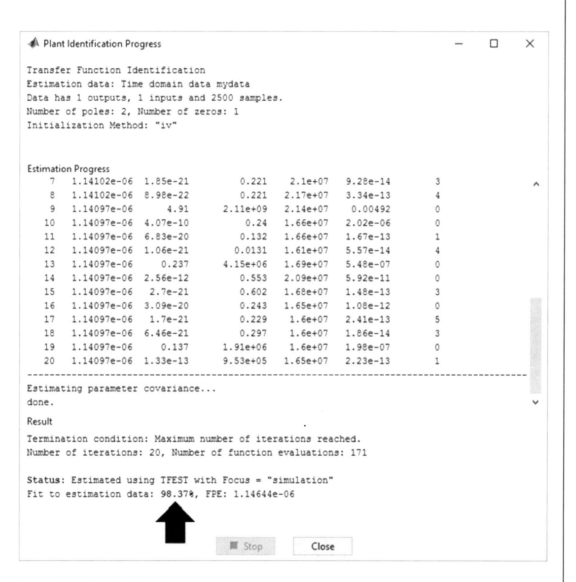

Figure 4.81: Result of analysis.

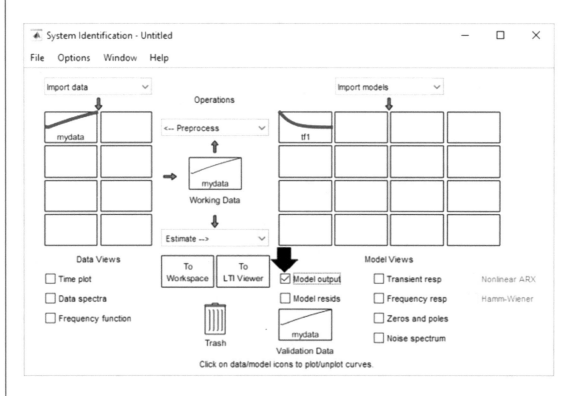

Figure 4.82: "Model output" check box.

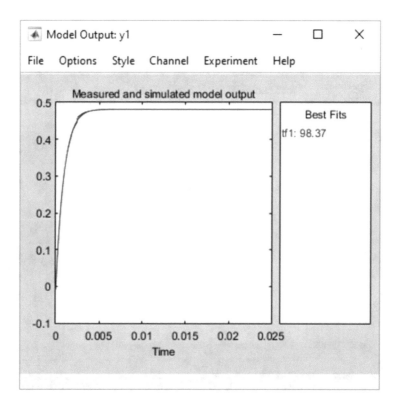

Figure 4.83: "Model output" window.

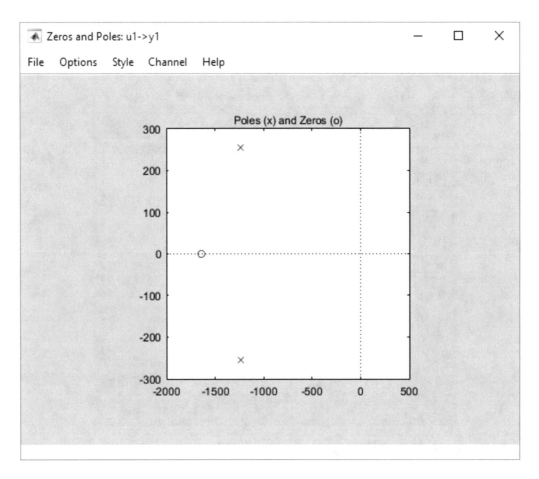

Figure 4.84: Pole-zero diagram of the estimated model.

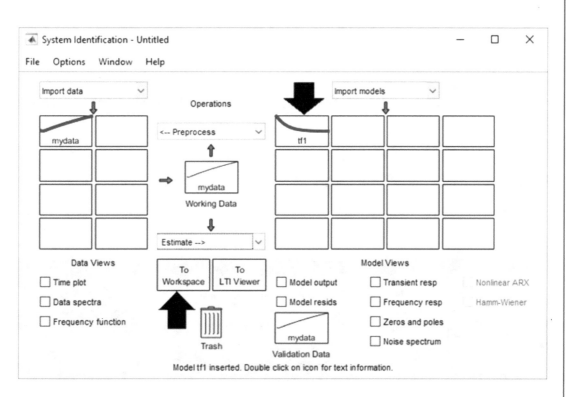

Figure 4.85: Exporting the estimated transfer function to workspace.

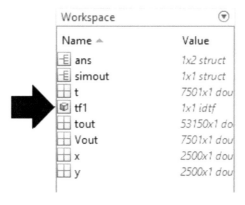

Figure 4.86: A variable named "tf1" is available in the workspace.

```
Command Window

>> t=simout.time;
>> Vout=simout.signals.values;
>> plot(t,Vout),grid on
>> x=t(find(t>.05))-.05;
>> y=simout.signals.values(find(t>.05))-simout.signals.values(find(t==.05));
>> plot(x,y),grid on
>> ident
fx >> pidTuner(tf1)
```

Figure 4.87: Running the PID Tuner application.

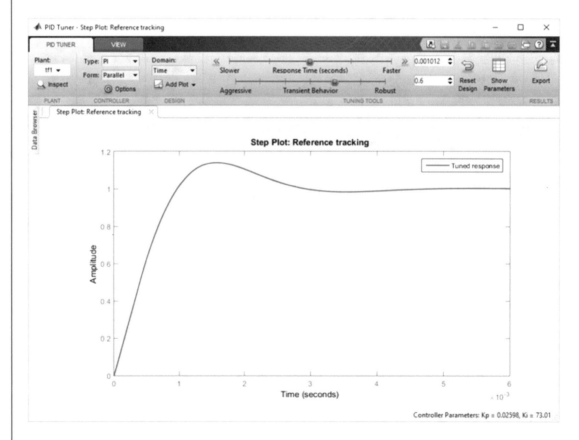

Figure 4.88: PID Tuner window.

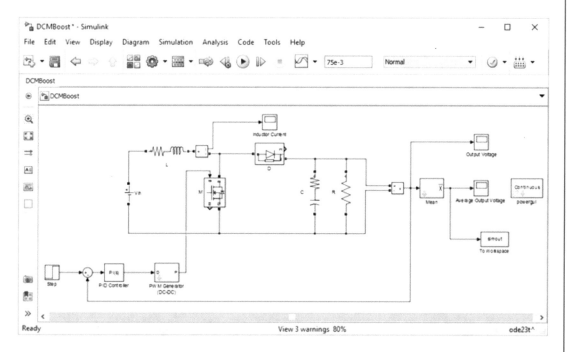

Figure 4.89: Closed-loop Boost converter.

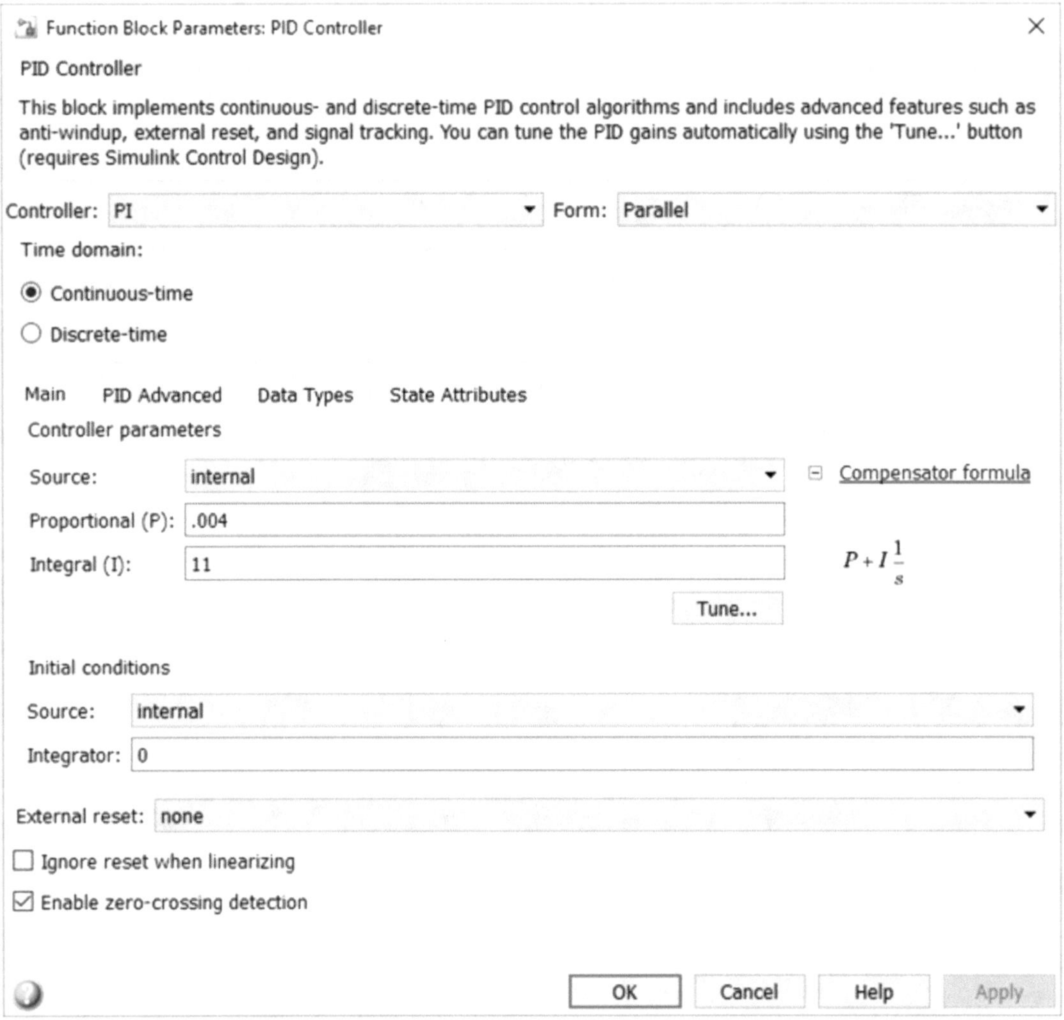

Figure 4.90: "PID Controller" block settings (see Fig. 4.89).

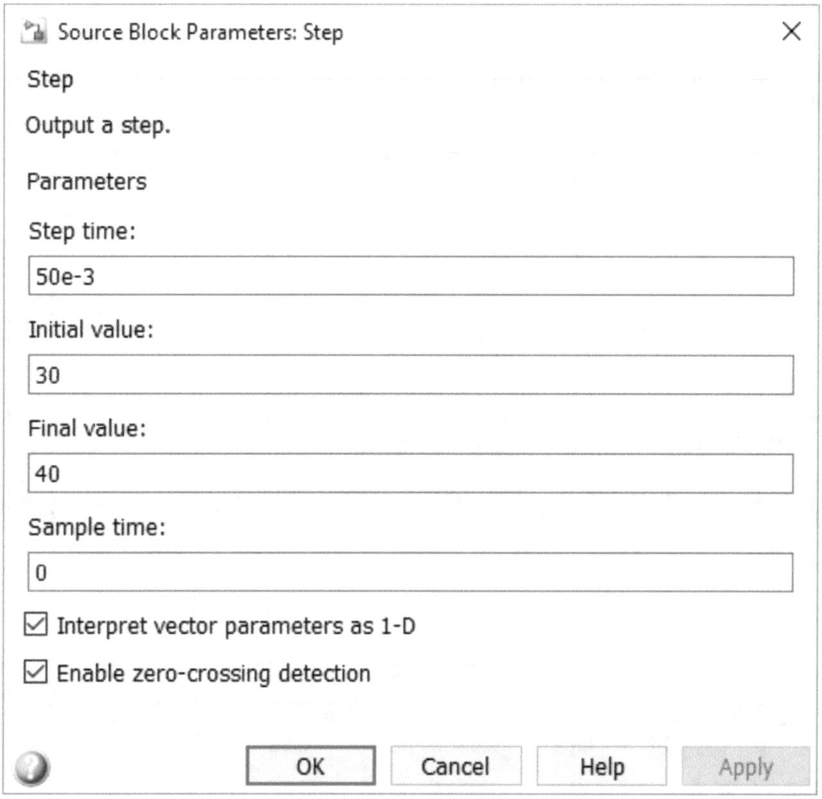

Figure 4.91: Step block settings (see Fig. 4.89).

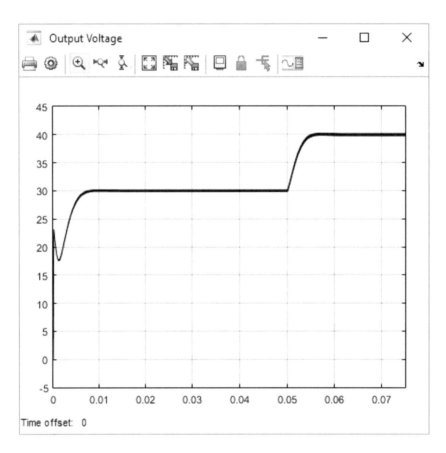

Figure 4.92: Simulation result (see Fig. 4.89).

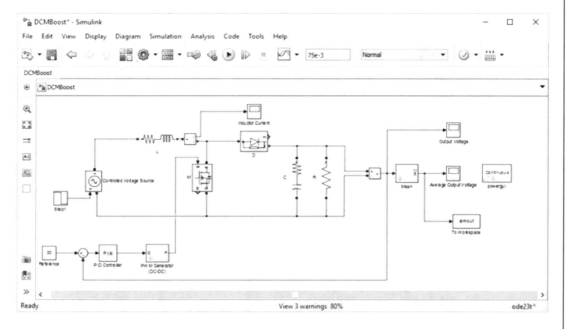

Figure 4.93: Simulation diagram to study the effect of change in input voltage.

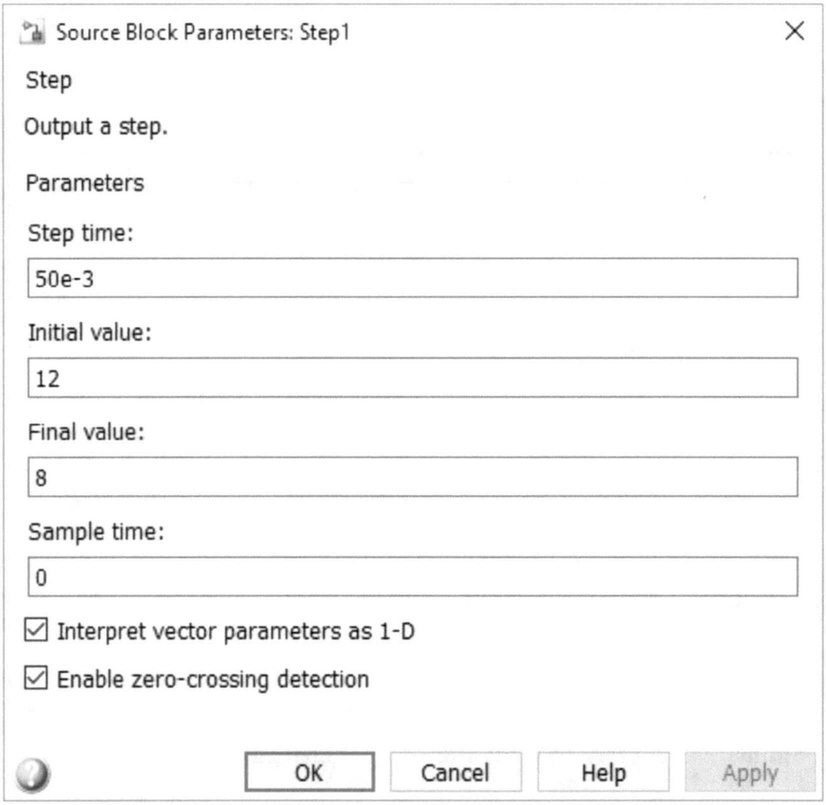

Figure 4.94: Step block settings (see Fig. 4.93).

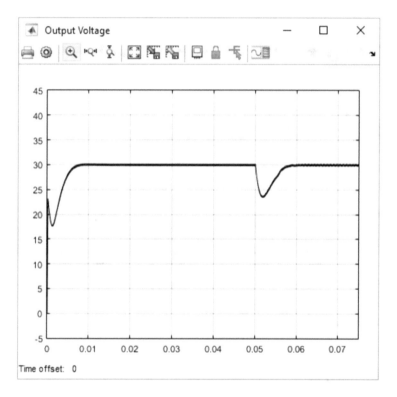

Figure 4.95: Simulation result (see Fig. 4.93).

As another scenario, load changes suddenly from 50 to 25 Ω at $t = 50$ ms. Input voltage is constant (12 V) in this test scenario. The simulation result is shown in Fig. 4.96. The controller keeps the output voltage constant despite load changes.

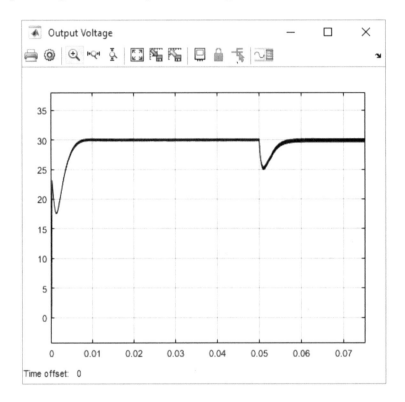

Figure 4.96: Simulation result.

4.7 CONCLUSION

In this chapter we studied controller design problem for DC-DC converters. We used the powerful MATLAB®toolboxes to design the controllers.

REFERENCES

[1] Daniel Hart, *Power Electronics*, McGraw Hill, 2011.

[2] Robert Erikson and Dragan Maksimovic, *Fundamentals of Power Electronics*, Springer, 2001. DOI: 10.1007/b100747.

[3] Simon Ang and Alejandro Oliva, *Power Switching Converters*, Taylor & Francis, 2005.

[4] Marian K. Kazimierczuck, *Pulse Width Modulated DC-DC Power Converters*, John Wiley, 2012. DOI: 10.1002/9780470694640.

[5] Christopher Basso, *Designing Control Loops for Linear and Switching Power Supplies*, Artech House, 2012.

[6] Seddik Bacha, Iulian Munteanu, and Antoneta Iluliana Bratcu, *Power Electronics Converters Modeling and Control*, Springer, 2014. DOI: 10.1007/978-1-4471-5478-5.

[7] H. Sira Ramirez and R. Silva Ortigoza, *Control Design Techniques in Power Electronics Devices*, Springer, 2006. DOI: 0.1007/1-84628-459-7.

[8] Dean Venable, The KFactor: A new mathematical tool for stability analysis and synthesis, *Proc. Powercon 10*, 1983.

[9] Katsuhiko Ogata, *Modern Control Engineering*, Pearson, 2009. DOI: 10.1115/1.3426465.

[10] Gene F. Franklin, J. Da Powell, and Abbas Emami Naeini, *Feedback Control of Synamic Systems*, Pearson, 2014.

[11] Chi-Tsong Chen, *Analog and Digital Control System Design: Transfer Function, State Space and Algebraic Methods*, Oxford University Press, 2006.

[12] Katsuhika Ogata, *State Space Analysis of Control Systems*, Prentice Hall, 1967.

[13] Chi-Tsong Chen, *Linear System Theory and Design*, Oxford University, 2012.

Authors' Biographies

FARZIN ASADI

Farzin Asadi received his B.Sc. degree in Electronics Engineering and his M.Sc. degree in Control Engineering and Ph.D. degree in Mechatronics Engineering. Currently, he is with the department of Mechatronics Engineering at the Kocaeli University, Kocaeli, Turkey.

Farzin has published 22 international papers and 2 books and is on the editorial board of 6 scientific journals. His research interests include switching converters, control theory, robust control of power electronics converters, and robotics.

KEI EGUCHI

Kei Eguchi received his B.E., M.E., and D.E. degrees from Kumamoto University, Kumamoto, Japan, in 1994, 1996, and 1999, respectively. Currently, he is a professor at the Fukuoka Institute of Technology, Fukuoka, Japan.

Kei is a president of the Intelligent Networks and Systems Society, an associate editor of *IJICIC*, an associate editor of *ICIC Express Letters*, and a senior member of IEE of Japan, APCBEES, IRED, and SAISE. He has published more than 200 international papers. His research interests include switching converters, nonlinear dynamical systems, and intelligent circuits and systems.

Printed in the United States
by Baker & Taylor Publisher Services